*Salomon Smith Barney
Retail Research Team*

A SHEARWATER BOOK

CONSUMING DESIRES

Consuming
DESIRES

Consumption, Culture, and
the Pursuit of Happiness

Edited by
Roger Rosenblatt

ISLAND PRESS / Shearwater Books
Washington, D.C. • *Covelo, California*

Library of Congress Cataloging-in-Publication Data
Consuming Desires : consumption, culture, and the pursuit of happiness
 / [edited by] Roger Rosenblatt
    p.   cm.
  Includes bibliographical references and index.
  ISBN 1–55963–535–5
   1. Consumption (Economics)—United States.   2. Consumer
behavior—United States.   3. United States—Social conditions—1980–
4. Individualism—United States.   5. Mass society.   I. Rosenblatt,
Roger.
HC110.C6C586   99–21667
306.3'0973—dc21   CIP

Contents

Consuming
DESIRES

Roger Rosenblatt

He took out a pile of shirts and began throwing them one by one before us, shirts of sheer linen and thick silk and fine flannel which lost their folds as they fell and covered the table in many-colored disarray. While we admired he brought more and the soft rich heap mounted higher—shirts with stripes and scrolls and plaids in coral and apple green and lavender and faint orange with monograms of Indian blue. Suddenly with a strained sound Daisy bent her head into the shirts and began to cry stormily.

"They're such beautiful shirts," she sobbed, her voice muffled in the thick folds. "It makes me sad because I've never seen such beautiful shirts before."

—F. Scott Fitzgerald, *The Great Gatsby*

It occupies barely a minute, that scene in *The Great Gatsby* in which Daisy and Nick Carraway watch their host heap his impressive hoard of shirts on the table before them, but in many ways it is the center of the book, the painful and thumping heart of the rapacious American appetite.

Gatsby tells his two guests that he does not buy the shirts himself; "a man in England" sends him selections of clothing at the outset of each spring and fall. Every season come the shirts in glorious abundance, in all colors and styles and fabrics, and of great expense—more shirts, we imagine, than any man could wear in a lifetime, yet never enough. Why does Daisy "cry stormily" when she sees them? Because she has a simpleminded lust for things, to be sure. But also, one may suspect, because as beautiful and plentiful as the shirts are, there can never be a sufficient number of them to satisfy her longing for God-knows-what. And she is in the minor leagues of desire compared with Gatsby himself, who is all longing, all yearning for things like shirts, like Daisy, in which he believes lies the secret of American happiness.

To have and have not: that is the essence of consumerism in America and the engine of Western capitalism—to experience a moment when the acquisition of what George Carlin calls "stuff" brings you to an epiphany of sobbing because its very being, and the act of acquiring itself, reminds you of the stuff you do not have, or do not have yet. The thrill of the present drags you into the want of the past and then propels you into prayers for the future. The stuff is there and not there.

It is a strange basis for a civilization, but an effective one. Directly and indirectly, some 90 percent of the American work force is in the business of producing consumer goods and services. Consumer products make up what we are. They generate most of the nation's income and employment. And they travel, bringing both themselves and the desire to have more to countries that have less and nothing, so that one fine day those places, too, by getting and spending, can have and have not.

So it is not that the acquisitive impulse is without benefits. Nor is it acted on without regret and spiritual reconsideration. In America, spiritual reconsideration is part of the process. We not only know how to have and have not; we also know how to—indeed, prefer to—have it both ways regarding the moral consequences of excessive appetites. In every era, especially one such as this, of uproarious economic prosperity, voices of lamentation offer guilt as balm. In the mid-1800s, Ralph Waldo Emerson and Henry David Thoreau gave us elegant scoldings. American economist Thorstein Veblen gave a name to our "conspicuous consumption" and berated our "predatory culture."[1] More recently, in February 1998, Hillary Rodham Clinton warned the World

Economic Forum in Davos, Switzerland, that "consumer capitalism" is undermining "the kind of work ethic [and] postponement
of gratification . . . historically associated with capitalism."[2] She
is not right on the history, but she is part of a very long tradition
of self-recrimination for all that has made America work, and for
which it apologizes.

Even our expressions of contrition have a double-edged effect.
We speak of "moral bankruptcy" and "emotional poverty" in criticizing the spending-spree culture, as if matters of the spirit, like
other commodities, were best understood in monetary terms.
Language reveals that we mean our apologies and that we do not.
How could it be otherwise? America did not transform itself into
history's most powerful civilization by abjuring material goods.

Yet regrets duly expressed and accounted for, it is the perpetual and relentless round of having and yearning that drives the
system, and which, in my view, maintains us in a continual state
of unhappiness, conscious or not, a state we may actually seek.
The state of not having gives us a frontier, a thing to reach and not
to reach. When we ran out of real frontier, we made other things
to not have. Other Western cultures grow rich, but they seem to
achieve contentment with their riches; they settle down when
they are up. Not us. Compare another brief literary scene with
the shirt spilling in *Gatsby*. In John Galsworthy's *The Man of
Property,* the hero, Swithin, bursts with pride over his magnificent
cut-glass dining-room chandelier because it forces others to perceive him as a man of wealth who "deserved a solid and prolonged
happiness."[3] In Dickens's *Oliver Twist,* "more" means merely an
adequate amount.

This is not to suggest that England or any other Western nation
has achieved a high state of moral being because it has learned to
rest more comfortably with its purchases. If America has been the
most consuming civilization, it has also been the most generous.
But the problem of the satisfied/unsatisfied appetite seems to be
peculiarly ours, and it is getting worse. Luxuries are deemed necessities.

A Neiman Marcus clothing and perfume catalog is titled
Essentials. Hammacher Schlemmer sells a Baby Elephant

Sprinkler Topiary ($249.95) that sprays water through its moss-covered trunk and Foot-Warming Spa Booties ($44.95) for use after a pedicure. Catalogs that sell everything from gems to dog blankets pile up at the door or in the mailbox as never before, thanks to the selling of debt on the part of credit card companies. Instant shopping malls pop up in every village and in the larger towns. My neighborhood in midtown Manhattan is packed with blazing new apartment houses, each rising from a ground floor of stores with familiar names. In a five-minute walk through my accidental urban mall, I pass Eddie Bauer, the Gap, Banana Republic, Gracious Home, Starbucks (two of them), an athletic-shoe store, a furniture store, purveyors of gourmet foods, and banks with automated teller machines. One vendor feeds the next.

There are things for sale that have never been for sale before. I cannot watch an NBA basketball game on television without seeing various products scroll down on a gizmo set up at courtside or without being told by a commentator that someone who makes a good pass in the game is responsible for an "AT&T Great Connection." The Sherwin-Williams Company, maker of Dutch Boy paints, sponsors memorable moments under the basket, "in the paint." The game is televised via a "Marriott Marquis Sky Cam," which sometimes captures a great assist, called a "Perry Ellis Look."

Politics is for sale, but not in the obvious ways that campaign contributions would suggest. Political office is sold by means of public opinion polls, which have turned the thoughts of a selected population into a commodity available to the candidates. The selling of any president these days depends on the quality of the public opinion he buys, and as in any market situation, the best-selling, though not necessarily the best, opinion wins.

In the week during which I wrote this introduction, a man in Minneapolis married a woman he had met only the day before, and for whom he had shopped among hundreds of competitors. The fellow said he had set a date, June 13, 1998, by which he promised his family he would be married. As the date approached, he advertised for a bride on the Internet and elsewhere and set up a system of friends who functioned as buyers, interviewing candidates as one would inspect cars or meat. The "winning" bride was perfectly willing to go along with this farce (she is a pharmacology student at the University of Minnesota, and on television she seemed anything but a bimbo), as were all the other candidates. In short, every one of these educated young people was happy to treat marriage as a sales transaction. The groom was in the market for a wife, and he (and she) strolled off with what the market would bear. Who was surprised that the wedding ceremony took place in the Mall of America?

Ordinary citizens are becoming like the big spenders of the past, going on near-desperate hunts for new stuff to buy and to long for. Americans used to amuse themselves by recounting the extravagant and wasteful doings of the rich—Commodore Cornelius Vanderbilt's ballroom replica of Versailles, for example, or his grandson's Newport "cottage," the Breakers, with its seventy rooms, its billiard hall of green marble, and its wrought-iron fence that cost $5,000 just to paint (in the 1920s). Potter Palmer, the Chicago millionaire store and hotel owner, bought so many diamonds for his wife, Bertha, that when she wore them all at once she could not stand upright. William Randolph Hearst built an entire Bavarian-style village because he felt like it. People used to laugh at that sort of behavior, in part because they envied it. But in these flush days everyone believes that he can be a millionaire himself, and the stock market is proving a lot of people right. Why mock the rich when you can join 'em?

With every purchase, nonetheless, there is always some other purchase just out of reach (not for nothing was the light on Daisy's dock green). Perhaps American consumerism is a reflection of other impossible yearnings that define the Republic— equality, for instance. To say that we are a people of impossible longings is not to excuse excess but merely to note a problem that is gaining in urgency and significance as rapidly as the economy. When we run out of things to have, will we also run out of things to want? Or will it be too late to save ourselves from the acquisitive impulse, along with the rest of the world that we infect with it? In terms of the "sixth great extinction crisis" Stephanie Mills

refers to in her essay in this volume, will we finally assert our pre-
dominance as a species by killing off all the rest, and ourselves in
the bargain?

The Mills essay and the others presented here are layered and
complicated, and I do not mean to reduce their complexity when
I cite two ideas that are basic to nearly all of them. Both ideas are
grounded in Western forms of democracy and in Western capitalism, both have connections with Western psychology, and both
have to do with having and having not.

The first idea concerns with desire. In her essay "Movies and
the Selling of Desire," Molly Haskell refers to the Freudian idea
that people are in a perpetual, often tragic, state of yearning from
the moment of separation from the mother. And it is yearning,
writes Edward N. Luttwak in "Consuming for Love," that makes
the poorest Americans, who have no savings and small incomes,
borrow themselves to death. Yearning drives consumption, and
one way to begin to grasp this massive, pleasurable, painful, and
finally destructive impulse is to understand simply that we yearn.

The word *desire* comes up regularly in these essays. Haskell
writes about how movies began by selling certain images of a
more gratifying style of life and of a more gratified self. Luttwak
sees consuming as an addictive substitute for family nurturing. In
"What's Wrong with Consumer Society?" Juliet Schor discusses
the "aspirational gap" that relates to the burden of free will. Alex
Kotlowitz, in "False Connections," describes the inner-city
teenagers who emulate their suburban counterparts, and vice
versa, in a sort of emotionally compensatory consumption. In
"Consuming Nature," Bill McKibben says that for us Americans,
"all things orbit . . . desires." Mills calls consumerism "soul
famine." In "One World of Consumers," William Greider refers to
the peculiarly American form of melancholy that results from the
purposes and effects of capitalism, namely, to keep gratification
just out of reach.

The second idea underpinning these pieces is the Western
concept of self, especially the solitary and restless self. Greider
notes: "The world will remain caught in a profound political stalemate on the environmental question until we Americans learn to

put aside empty self-righteousness and accept the full burden of our historical position," which means an exaggerated power given to the self as well as "triumphalism." As the populous Eastern societies begin to emulate Western consumerism, it is scary to think that the seeds of this emphasis on unfettered individual satisfaction could grow and infect cultures whose psychological orientation is much more conducive to working together and to working with what one has.

Individualism and desire, then, are what make us great and what make us small. Freedom is our dream and our enemy. The essays touch on these paradoxes, and though all are too nuanced and graceful to preach easy reform, they give an idea of what reform means, where it is possible, and, in some cases, where it may not be as desirable as it appears.

Writing on the consumption of literature in "When We Devoured Books," André Schiffrin demonstrates that a broad choice of serious work, widely available and widely read in the nineteenth century, is now increasingly hard to find. In the years following World War II, he says, editors at one of the biggest mass paperback houses "didn't see the readers as segregated into an elite public and a mass public to which one had to pander"; their primary concern was "how to present new and very demanding work to a mass audience." The takeover of Random House by the Newhouse corporate managers was the first indication that cultural and intellectual values were being replaced by market values. Family businesses were becoming large corporate holdings, and whereas it had once been the slow growth of the firm as a whole that had come first, now it was the annual profit, a shift in priorities that basically pressured publishers to quintuple their sales figures. A logical system put in place throughout the industry "became a kind of iron mask that allowed for very little variation." Sectors were broken out into separate "profit centers," and "gradually, the pressure was put on to have each book pay its own way." The recent sale of Random House to a foreign conglomerate exemplifies completion of the process of commodification.

Books, says Schiffrin, have become like any dry goods on shelves. "A form of market censorship has firmly established itself," he writes. "The decision to take on a book . . . is now basically made according to the simplest of criteria. . . . Not is it interesting or important but is it hot, commercial, popular?" The greatest danger here, he warns, is to free thinking and ideas, including subversive ideas. If books compete as do other commodities, victory will go to the most convenient and to the cheapest. It remains to be seen whether American society will be informed of new and dissenting ideas when large publishers no longer play a crucial role in "disseminating new ideas to the broadest of publics."

Yet the public, too, is losing security in the bargain by buying itself into peril and poverty. In "Consuming for Love," Edward Luttwak contends that the structural changes in American society and the concomitant personal economic insecurities wrought by the highly vaunted efficiency of today's "turbo-capitalism" have resulted in a species not yet adapted to living without the soul-affirming "hugs" of an extended family. Thus, "most Americans are emotional destitutes, as poor in their family connections as are Afghans or Sudanese in money." As a substitute, and a way to cope with the resulting anger and insecurity, all but (ironically) the poorest 20 percent of American households buy themselves objectless and functionless presents in inordinate amounts, saving less than nothing and "borrowing with abandon from all possible sources." This creates a paradoxical and vicious cycle within which they sacrifice "their personal freedom and family life just to consume more."

Movies, our most powerful cultural medium, abet this tendency, writes Molly Haskell in "Movies and the Selling of Desire." Just as movie moguls once appealed to the desires of disenfranchised immigrants who wanted to emulate the Wasp elite, movies now are attempting to seduce "a lowest-common-denominator mass of minimal literacy and maximal impressionability and buying power"—American teenagers and a global audience of action movie fans. "From their inception, movies—both pusher and product—fed, paralleled, and to some extent created consumerism" by homing in on unconscious needs and cravings. What began as a unifying force, consumerism and movies—in the pattern of capitalism itself—"mutated into avatars of movement and change." Looking and buying became intertwined as "the notion of appearances . . . gradually came to have less to do with character and reputation and more to do with simply looking good."

Until the 1970s, the ideal viewer/consumer was female, and her primary vocation was to make herself into an object of desire, the "outer envelope" being a metaphor for inner beauty. In providing an outlet and a language for aspiration, "movies have played a role both progressive and conservative, democratizing in their populist appeal, conformist in their acceptance of the status quo, and paradoxical in the built-in, ongoing conflict between elitism (the stars) and egalitarianism (the story)." Just as James Bond films capture the artifacts and attitudes of late-twentieth-century consumerism, movies both reflect and create social conditions, but "their special charm is to offer fantasy clothed as virtual reality, a world where people consume without the tedium of labor."

On the same track, in "What's Wrong with Consumer Society?" Juliet Schor explains the "new consumerism," in which "competitive consumption" feeds a system of widening class distinctions wherein people aspire to the level of consumption that others have achieved. The system focuses on visible, brand-name goods rather than more beneficial alternatives, such as leisure or savings. "More and more, what you wear and what you don't wear define who you are and where you are located on the social map." As "the Joneses" have given way to the work place as a social reference point and as face-to-face socializing has been replaced by hours spent in front of the television set (which presents a skewed picture of spending patterns), the gap between what people wish for and their actual income (the "aspirational gap") has widened to an amount that is "currently more than twice the national median household income." We are bound to experience "a profound failure at the heart of the global economy," says Schor, if human beings cannot adequately control themselves in an unconstrained consumer society.

"It all begins with housework," says Jane Smiley. "The way we live our lives today reflects what our ancestors aspired to get away from." Feminism and consumerism are strongly linked because, consistent with American ideals of freedom and individuality, capitalism responded to the enslavement of nineteenth-century women to domestic labor (dangerous cooking, constant fire tending, and hauling of water, not to mention cleaning, voluminous laundry, and child care) by inventing appliances, electricity, supermarkets, cars, schools, telephones, and running water—a level of expensive consumerism American women today would be hard-pressed to give up. "There is much talk of the emptiness of modern life, but think of emptying chamber pots of the accumulated waste products of seven or eight household members every day for the rest of your life. Think of asking someone else to do it. Think of shirking your duty and not doing it." But "the world's 5 billion or more people cannot live as Americans aspire to live." If they try, asks Smiley, how will we all survive? We need a new form of capitalism—indeed, a new society—to address this crisis.

We need to get away from making "false connections," writes Alex Kotlowitz. It is as consumers of certain expensive, highly visible, brand-name goods that the poor, socially and spiritually isolated inner-city kids who live in the ruins of Chicago's West Side housing projects participate in the secure and prosperous world of white suburbia. Companies such as Nike, Inc. (and even the makers of Hush Puppies, now redesigned in "gotta-look-at-me" colors) gear their advertising to these children, who, as both consumers and drivers of inner-city "style," feel comforted by this meager amount of control over their lives and imagine themselves to be connected to the wider society. Those same white teens from whom the urban poor kids imagine they take their fashion talismans mimic the urban "beltless, pants-falling-off-hips" fashion of their emulators, imagining ghetto life to be the sort of edgy, gutsy, and risky adventure so attractive to teenagers. "And so, in lieu of building real connections—by providing opportunities or rebuilding communities—we have found some common ground as purchasers of each other's trademarks."

And since so much of the consumer impulse is carried and re-inforced by the news media, the press itself must recognize the consequences of having become like a store. We need "a news consumer's bill of rights," says Suzanne Braun Levine. Citizens (seen as consumers) and the press (seen as mad dogs) must "come to an informed understanding of which parts of the problem" regarding the news (seen as product) and how it is perceived "are a function of press performance . . . and which are the result of a host of other societal and economic factors." This understanding is hard to achieve in the current environment of Internet misinformation, global media conglomerate influence, and the chronic concern of all news entities regarding the bottom line. "There is evidence that as the media have been absorbed into the big-stakes business world, money has become a more compelling interest than the public's right to know."

A news consumer's bill of rights—notwithstanding (or particularly considering) the many difficulties and dilemmas involved in news gathering and reporting—would require members of the press to be fearless and compassionate; to gather all relevant facts and assemble them into the truth; to be fair, objective, and unbiased; to inform and to be meaningful and responsible; to not be motivated by self-interest or commercial interest; and to not incite violence or invade individual privacy. If "the primary mission of news reporting were to be *fair to the public*," crime stories would not take precedence over most others, for example; complex events would be given more than mere headline treatment; and news would not be defined in terms of simple opposing forces. "When thinking isn't part of what the news is selling," Levine points out, "the public is not being fairly served." Journalists and the public must work together to "co-opt the crass brand-consumer relationship and focus on reforming the very system that is profiting from it."

How do we begin to get hold of the necessity of moderation and restraint? Stephanie Mills tries good, old-fashioned fear. "Can't get that extinction crisis out of my mind," she writes. The fact is, "hungry people will gorge themselves. It may be that today's rampant consumerism betokens the soul famine of a society estranged from the living earth." The earth's sixth great extinction crisis—wherein species are being lost at about a thou-

sand times the normal rate—is the result of humanity's vastly inordinate impact on the planet. The destruction of habitat began
with agriculture and then was accelerated by the power elite and
by commerce. Now the culprit is global consumerism. The sustaining of life has become increasingly contingent on people's
ability to pay for, rather than raise or make, the basic necessities.
Today, "because the sources, processing, and manufacture of our
goods are so widely scattered, it's nearly impossible for us to comprehend the effects of our way of life on the biotic community."
If we do not return to a decent relationship with the land, including each place's whole community of life—"furred, feathered, finned, fanged, and fungal," even the most parsimonious
lifestyles will become unsustainable.

David Orr offers the same thought, though within a more
serene framework. In "The Ecology of Giving and Consuming,"
Orr describes a letter opener he received as a gift from a master
wood craftsman as "an embodiment of skill, design intelligence,
kindness, and thrift," qualities needed to rechart a course away
from compulsive, excessive consumption and toward an approach
that leads to "ecological competence, technological elegance, and
spiritual depth." Appropriate ecological design aims to increase
local resilience by building connections among people, between
people and their places, and between people and their history;
"takes time seriously by placing limits on the velocity of materials, transportation, money, and information"; "eliminates the concept of waste"; and "has to do with system structure, not the coefficients of change." In short, "design at all scales entails not just
the making of things; it becomes, rather, the larger artistry of
making things that fit within their ecological, social, and historical context."

But Bill McKibben reminds us of the value of human modesty
in this process. He shows that "consuming nature" is a complex
idea and not without ironies. For a few weeks every spring, residents of Johnsburg, New York, are tormented by clouds of vora-
cious blackflies. The townspeople came up with a plan to treat
the streams with a naturally occurring bacterium that kills blackfly larvae, which appealed equally to environmentalists and own-

ers of local businesses. After all, it seemed perfectly reasonable to want to have air that is free of irritating insects. Since they "want to consume bite-free air," why not convert "the natural into the convenient"? After all, the unspoken motto of the consumer society is *"You are the most important thing on earth. . . . All things orbit your desires."*

McKibben meditates on his own opposition to the plan, humbly allowing that his nonconsumption of this convenience—his pleasurable sense of superiority in meeting the challenges of "the Rugged Frontier Adirondacks"—is no more than an ironic expression of image building through consuming: "Instead of defining ourselves by what we buy, we define ourselves by what we throw away." Still, this is a benign form of consumption and one that might lead to our being "exposed to forces that might actually change us" and that might remind us that we really *aren't* the center of the world.

In "Oh, Isaac, Oh, Bernard, Oh, Mohan," Bharati Mukherjee demonstrates that taken with a different attitude, even consumption may be tamed in civilization. The successful South Asian owner of an ethnic fast-food place in 1978 Toronto appeared, at first glance, to be an example of the immigrant philosophy that "if you treated the New World right, it would make room for you at its table of conspicuous consumption." But this community-based prosperity fed hostilities, and the stances adopted by the various parties in a public debate of the issues revealed to Mukherjee that "every North American city was becoming freshly a frontier" with an "insistent us-versus-them mode of self-defense."

Until 1920, most South Asian immigrants to North America were uneducated and did not speak English. Scapegoated as "an unassimilable alien group of widow burners and sneaky mystics," they ended up as "familyless drifters through the American dream–scape." The second wave of immigrants benefited from having superior educations and the good timing to arrive during the 1960s, when Asian Indians were credited "with having automatic access to wisdom and serenity. . . . Even more than being eager acquisitors, they were dedicated family men." In the new country, the family unit shrank from extended to nuclear, which made the "quality of family life less expensive to enhance." In the 1970s and 1980s, these "smart consumers of the American class system" "drove in the American lane at work and in the Indian lane at home." Their children, though, are rejecting those "filial examples of ambition and diligence" and "looking to self-empow-ered minority communities . . . for role models," remaking "the drama of Americanization." To South Asian housewives, the mall is no consumerism nightmare. Rather, to be unfettered by tradi-tional social constraints, "to have friends who will waste a morn-ing with you in public, is to be free."

Martin E. Marty gives a similar view macrocosmically. In "Equi-poise," he asks: Whose morals and whose ethics will come into play regarding consumerism and consumption issues in the newly triumphant global economy? In America, "approved social con-tracts," which are deposits and experiences from the past, have been formed and informed by three relatively incompatible philosophies: from the late eighteenth century, the biblical her-itage and the Enlightenment heritage, and from the late nine-teenth century, "social Darwinism." This society, like its individu-als, lives with the conflicting sets of claims of the religionists and the free-enterprisers. Those from the "Gospel traditions" should know that "to disdain what is on earth to be consumed is not purely and simply virtuous."

Only the resolve of the individual, says Marty, can lead to a bal-anced and positive practice of consumerism. This is not to be achieved by seeking "banal middles" but rather through cultivat-ing equipoise (quoting John C. Haughey, S.J.), "a poise of spirit that can weigh conflicting pulls and not act compulsively or ad-dictively so that one's choices or use of things express one's deeper self, one's interiority."

William Greider puts it all in an environmental perspective in "One World of Consumers." From Burger King restaurants in Malaysia offering Islamic dressing to a Buddhist temple in Bang-kok festooned with discarded plastic bags, "the action in the de-veloping countries is like a loop of old film that continuously plays back our own history for us" as the aspiring poor mimic the

American prosperity based on industrialization and mass consumption. "If the world is to save itself from ecological disaster, the redemption cannot begin among the poor, however satisfying that idea might be for the missionaries. Only the wealthy few— that is, nations such as ours—have the power and the wherewithal to rescue us all from the impending consequences of mass consumption on a global scale." Economic inequality is fundamentally an environmental issue, and rising incomes and consumption are necessary elements of a solution. Growth must be redefined in qualitative rather than quantitative terms, consumer expectations must change, and there must be "a new social understanding that like the global system itself, we are all in this together now and no one will be saved unless all of us are."

Greider's message requires a revolution of thought that Americans may not be willing to make. The triumph of the West over the Soviet Union in the 1980s was at least in part read as the triumph of capitalism over socialism (though that is not what it was), and thus the very idea of a cooperative society ran the risk of being tainted as being un-American. Americans always strive for equality—Tocqueville understood that we are interested more in equality than in freedom—but when at the same time we strive for separation from one another, all that freedom achieves is isolation. In several of the essays presented here, the 1970s are mentioned as the time when rampant consumerism and mass consumption began to reach a crisis point. That crisis has grown nearer in direct proportion to the increase in isolation in the culture, due largely and most recently to the computer and other emerging technologies (which also came to prominence in the 1970s) that have driven us farther both from reality (place is now virtual place) and from one another.

All American inventions are created with a mind to obliterate the class system, and all wind up reaffirming the various divisions or creating new ones. First, a few people have a car, a telephone, a television set; then, a great many have them. Yet nothing is really changed by these acquisitions, and the class system endures because of other circumstances created by the new inventions. Americans do not like to admit the existence of social

classes; we founded our nation as a retort to class-ridden Europe.
But our answer to Europe's class system was to set up—as an
ideal—a prosperous, landed Jeffersonian class to which everyone
would belong. What we have evolved into is a multiplicity of
classes, including the information class created by the emerging
technologies. The way America has restructured itself, people live
surrounded only by people who live as they do. Homes in like
neighborhoods cost the same; schools are alike; clothing and
manners are similar; people socialize with economic peers.
Americans live cocooned with members of their own class yet still
deny the idea of class in America. Indeed, so eager are Americans
to hide the idea from one another, they do not have the slightest
notion of what class they belong to.

The top 20 percent of American families make as much money
as the remaining 80 percent. The top 5 percent of that 20 percent
make as much as the remaining 15 percent. In that 5 percent,
once again the top one-fifth (that is, 1 percent of America) makes
as much as the other 4 percent. Yet if one were to ask a couple
making $40,000 per year before taxes and a couple making a pre-
tax $200,000 what class they were in, both would answer (hon-
estly and persuasively) not only that they belong to the middle
class but also that they are just scraping by.

There are at least two reasons for this status blindness, both
having cropped up since the early 1960s. One is the enormous
fissure that began to develop within the middle class between the
upper and lower extremes, and the other is the splitting of the
middle class into various subclasses. The 1980s (in spite of what
was said about "morning in America") was the time when the bottom began to fall out of the middle class. From 1960 to 1974, real
wages and salaries for workers increased by 20 percent. From
1977 to 1989, the wages of workers with less than a high school
education fell by 20 percent and the overall median wage for men
plummeted. Among full-time working men aged eighteen to
twenty-four, the proportion earning less than $12,195 per year (in
1990 dollars) more than doubled in the 1980s, from 18 percent
to 40 percent. Even the top third of the upper middle class has
been slipping fast. From 1989 to 1993, white-collar executive
wages dropped by 0.8 percent; technical workers' wages, by 2.9

percent; and those of college-educated employees, by 2.5 percent. In 1993, for the first time in history, there were more job-
less white-collar workers than blue. This economic rejiggering has
created classes within classes within classes—like Russian dolls,
without the smiles.

The emerging technologies are imposing a new class system on
the existing ones. The technological class system has as its overarching context the ability to use the new machines. If everyone
has a computer, everyone theoretically belongs to the same class.
But within that overarching class lie thousands of subclasses,
from chess players to militia members. The idea reinforced by
these separations is that everyone is on his or her own, that life
consists of goods acquired for oneself—indeed, that it is to be
best lived in splendid isolation, surrounded by as much stuff as
one can afford or, as Luttwak suggests, not afford.

To be sure, there are countersigns to this tendency. The new
Hall of Biodiversity in the American Museum of Natural History
is almost subversive in its stated purpose, that unless human beings learn to cooperate with other species, everyone on earth will
vanish. In a way, the Hall of Biodiversity stands as a fifth column
at the center of ambitious Manhattan and serves as a counter-
weight to the individual appetites that make up America's symbolic island. In its diorama of a portion of the central African rain
forest; in its assemblage of more than 1,000 examples of species
mounted along a 100-foot wall, like all of biology's general store;
in its electronic bio–bulletin board that views up-to-date news of
the planet, such as weather reports, the exhibit makes the clear
point that unbridled individualistic actions on the part of the pre-
dominant species—us—will end in the sixth mass extinction to
which Mills refers in her essay here.

The Hall of Biodiversity is not the only sign of a growing awareness that getting and spending have gone dangerously far. Books
are appearing that sound the alarm (though one is reminded of
Schiffrin's lament that critical, counterculture books, too, are an
endangered species), such as Yiannis Gabriel and Tim Lang's *The
Unmanageable Consumer* and Jackson Lears's *Fables of Abundance*.[4] The worrisome aspect of these works of criticism is that
others in the past, equally persuasive in the abstract, have gone

without affecting much but conversations among intellectuals. The road to environmental hell is paved with the good-intentioned efforts of Vance Packard, John Kenneth Galbraith, Guy Debord, Dorothy Davis, Jessica Mitford, Raymond Williams, and Marshall McLuhan.

As long as people believe that society relies on the cultivation of self at the expense of others, none of the instruments of warning stands much of a chance of success. Nothing seems to cultivate the self as does consumption; all one needs to do, in a moment of insecure self-inspection, is take an inventory of one's possessions to assure oneself that a worthwhile, substantial being has been constructed. To fortify this illusion, one may convince oneself that in the process of acquiring things, one is participating in, and touching the soul of, democracy. America is a gigantic supermarket, and the consuming citizen is keeping it healthy by acting on free choice. It goes without saying that a sense of inexhaustible choice must be what makes America tick.

Of course, the effect of this sort of thinking on America itself is the placement of an unbearable weight on every structure. In his recent book *Visions of the Future,* Robert Heilbroner, writing of the ways in which the pursuit of money drives history, says that "the ubiquitous effort to accumulate capital introduces a tremendous economic pressure that spreads through the system," from vast institutions to the lowly individual buyer.[5] If to be ever bigger and richer, and to own more and more, is the sole ambition of a civilization—or even if it is not, but none other is openly expressed—anyone who has doubts about the virtues of consumerism is deemed downright unpatriotic.

But unless one is content to stroll among the things of this world forever and to keep adding to them, wasting resources and awaiting the inevitable end, there has to come a point at which the constructed self gives way to the authentic self, as Martin Marty suggests. Lears's *Fables of Abundance* tells a spellbinding tale of how advertising men at the turn of the century decided to disassemble the Protestant work ethic and erect in its place a new "therapeutic order" that encouraged a cultivation of the self through goods and services designed to meet (or foster) psychological needs. At a time like the present, when communities are virtual if they exist at all, the individual constructs a self out of those needs. And yet he or she also perceives, if dimly, the effort as counterfeit. Otherwise, all that relentless yearning would not follow on the heels of all that relentless getting and spending.

One way to relieve the guilt of avid accumulation has been to separate production and consumption. Theoretically, Americans can consume guilt-free if they do not actually see the sweatshops in Asia or in Latin America that produce their dresses and running shoes. Because of the public nature of her television show, Kathie Lee Gifford's shame at the revelation that her line of clothing was created in sweatshops was made to seem merely one more buffoonish moment in popular culture. It was, in fact, an essential discovery for a lot more people than Kathie Lee that they were happy to consume without guilt by pretending to be spectators in life, not responsible participants.

To return to the scene of Gatsby's shirts, there is an earlier work of literature that deals with shirts from the opposite end. In the 1840s, at the outset of England's industrial revolution, poet Thomas Hood wrote "The Song of the Shirt." It was in response to an incident in which a poor widow with two infant children was brought to trial for pawning clothing belonging to her employer in a clothing factory. The trial exposed the wretched pay and living circumstances of factory workers and aroused the public's sympathy, as well as Hood's. He wrote:

> With fingers weary and worn,
>   With eyelids heavy and red,
> A woman sat in unwomanly rags,
>   Plying her needle and thread —
> Stitch! stitch! stitch!
>   In poverty, hunger, and dirt,
> And still with a voice of dolorous pitch,
>   Would that its tone could reach the Rich! —
> She sang this "Song of the Shirt!"

(Thomas Hood, "The Song of the Shirt")

Today that woman is Mexican or Chinese, but the basic circumstance remains unchanged, even if the public is more aware of it than it lets on.

In the long run, though, I believe that none of these delusions takes, that we are eventually left with the truth of our situation. As the essays here make clear, that truth is not simple. People like to consume. Indeed, consumption is often grounded in cultural legacies that wend their way back through genealogies to the whole meaning of family. Why should someone in America or anywhere else, who has struggled a lifetime to achieve the wherewithal to get and spend, not revel in the fruits of his or her labor? What other evidence does one have of achievement or progress?

Yet that, as all these essays suggest between, and in, the lines, is not necessarily a rhetorical question. There is other evidence of accomplishment and of existence, but a special exertion of the imagination is required to grasp it. The consumer society is based on want, and want is derived from visible things. The entire advertising industry was formed by clever arrangements and manipulations of stuff placed before our eyes. Yet the same people who lust after cars or car factories also readily, often balefully admit that the best stuff in their lives consists of invisibilities—"things invisible to see," as John Donne put it. Love, friendship, honor are to be culled from the air, and what proof one has of their existence comes solely from unseen feelings. The absence of evidence, as was said in the O. J. Simpson trial, is not the evidence of absence.

The trick—the essential, monumental, soul-healing trick—is to think of one's life as consisting of both real and imagined estate and to acknowledge that one's most valuable property is not, and never was, for sale. That, of course, has been the plea of all moralists since people started behaving badly. It is driven today by the practical power of dire consequences, and the matter is urgent. What we need is a search of the self, which, I believe, will reveal tendencies that run counter to all visible evidence.

In some way, everyone is Daisy Buchanan in tears, and Nick, and Gatsby, and all who stand helpless before the things they long for. Even the woman of Thomas Hood's poem dreams of the shirt.

The poem prays that her song will reach the Rich. It does. Rich and poor are connected by the same song, by the same shirts, and both dissolve in sorrow for something they do not comprehend. We are in a perpetual state of yearning, but the yearning may be for less, not more, for a simplification of existence that allows the poor to rise, the rich to level off, the other species to survive, and the world to go on. But we lack the knowledge or the will to achieve all that, and so we want and do not want, and weep without knowing it.

One World of Consumers

William Greider

MODERN LIFE has played a monstrous joke on the innocent American traveler who heads off to remote places in search of the exotic. When we tour the world these days, there is still the surprise and delight of glimpsing the strange ways of the "other." But we also can no longer avoid an awkward confrontation with ourselves or, rather, with the artifacts of our own civilization.

Ten thousand villages in China may lack streetlights, but they now glow at night with the soft light of television screens emanating from open doorways and windows. I went to one of those villages in Shaanxi Province to tour a military-run factory that manufactures advanced parts for commercial aircraft. The deep strangeness of the place made a mess of my inherited cultural presumptions.

The Hongyuan Aviation Forging & Casting Industry Company had a few modern machine tools imported from Germany and Japan, but its foundry barn looked crude and soiled in comparison with ours. A line of young machinists in blue smocks stood at their lathes in the factory gloom, looking like characters in a sepia photograph from the early industrial life of America—Detroit, 1920, or maybe Chicago, 1890.

The managers led me to the display room to view the various components—gear wheels, rings, rods, axles—that the company fashions for Chinese aircraft or exports as parts for steam turbines and other machinery around the world. The place of honor was reserved for five titanium-alloy struts that Hongyuan manufactures for the Boeing Company. Spread out on a blue velvet drape, these super-strength objects will support jet engines on Boeing's 747.

Okay, I thought, this is weird. The next time I'm flying on a 747, I will think of this poor but prospering village in China, where some of the people still live in caves dug into the valley wall and where the foundry's own testing labs are still hidden underground (Mao Tse-tung's quaint attempt to protect China's heavy industry from U.S. or Soviet nuclear attack).

The next morning, I walked through the market at the center of the village, where peasant farmers were lined up along high brick walls, squatting back on their heels behind mounds of carrots, greens, cauliflowers, scallions, and cabbages. One of the peddlers, a young man in a blue sweater, jabbered at passing housewives through a small battery-powered loudspeaker. The milky white powder he was selling was lotus flour. As I watched, he weighed out portions of flour on his handheld scale and poured them into clear plastic bags. He knotted each bag with the same deft twist of the wrist that I've seen the checkers use at my supermarket back home.

Globalization means, among other things, that there is really no place for us to hide from freighted connections with our own daily lifestyle. The profusion of "stuff" seems to be everywhere, including stuff that used to be exclusively for the well-to-do of the world. When I visited China and points beyond, I was traveling with a professional purpose, working on a book about the global industrial revolution and its economic and social implications. But the tourist in me was naturally drawn to the sheer wonder of artifacts of everyday American life creeping around the globe, showing up in the most unlikely backwaters.

At each moment of encounter, I experienced contradictory re-actions—the reflexive delight of recognition followed by a rush of foreboding. These responses are probably commonplace now for most tourists. What did *you* see in Malaysia? I saw teenagers with Walkman portable stereos and stacks of cheap CDs along the streets. I bought fake Rolex watches at the Chinese market. I saw Burger King restaurants offering Islamic dressing, and I drank root beer and heard rap music with lyrics in the local language, Bahasa Indonesia.

Some of the connections, of course, just seem loopy, in the very ways some self-important Americans have become ridiculous with-out realizing it. At a second-rate restaurant in Jakarta, I was dining alone and watching four Korean businessmen, perhaps factory managers or salesmen, at the next table. Each had a cell phone on the table, and while the men ate and conversed, each made oc-casional phone calls. Business is business, everywhere in the world.

I also glimpsed hints of tragedy, of a dreadful reckoning that is coming. In Bangkok, the golden temples are surrounded now by frantic commerce. The gentle and beautiful are forced to retreat before an onslaught of the quick and modern as the past is grad-ually obliterated by the new. The city's traffic jams are the worst in Asia, people will tell you with a mixture of frustration and pride. Bangkok's ancient canals are silted up, the water table is falling, and salinity is rising in the Chao Phraya River. In the countryside, it is probably worse. Monsoon rainfalls are weaker now, everyone agrees, because of the industrialized, agricultural development. Mangrove swamps along the coast are drained to grow shrimp for the sushi bars of Tokyo.

In a working-class neighborhood of Bangkok that I visited to in-

terview textile workers, I saw an especially poignant tableau outside the union's offices. At the corner of a vacant lot, the neighbors had erected a modest shrine—a miniaturized Buddhist temple set atop a pole—where passersby would say prayers and leave
humble offerings. Around the base of the shrine, the ground was
clotted with bits of plastic. Thousands of blue-and-pink plastic
bags, the tissue-thin shopping sacks that are ubiquitous in Asia,
blew randomly across the empty lot, accumulating evidence of a
higher civilization.

Thailand is catching up. The National Petrochemical Company in 1994 announced in its annual report that Thais now
consume forty-four pounds of plastic per capita each year. This is
a big improvement, the company said, but the country still lags
behind more advanced neighbors such as Korea, Japan, and
Taiwan. The Taiwanese consume more plastic per capita than
Americans do.

On learning the larger story of industrialization in developing
countries, the brutal inequities and social upheavals, the process
of natural degradation that accompanies the creation of wealth,
any sentient being yearns to cry out: "Wait! Stop! Don't you realize what you are destroying?" This is a natural response, perhaps,
but it is also, I think, quite arrogant—especially if one comes
from America, the world capital of mass consumption.

After all, these folks merely aspire to emulate what they inter-
pret as the American style—the American system of prosperity—
with some local elaborations thrown into the product mix. Why
would people choose to turn away from what so plainly succeeds?

In Kuala Lumpur, I talked with a Muslim intellectual, Merrill
Wynn Davies, a woman who was born and raised in Wales and
university educated in Great Britain. Knowing the social realities
back home, Merrill is contemptuous of Anglo-Saxon presumptions of law and justice. She is similarly impatient with Western
environmentalists who lecture poor countries on the evils of industrialization. She brushed aside my questions with this remark:
"What these people want is what the West already has. And why
shouldn't they? It is a very nice life, isn't it?"

Everywhere I traveled, from Asia to eastern Europe, I kept hearing her point repeated in various expressions of desire. Of course, the aspiring poor are bound to mimic those who are well off. Why shouldn't they? This simple truth, I have come to believe, now lies at the heart of the environmental question, like it or not.

If the world is to save itself from ecological disaster, the redemption cannot begin among the poor, however satisfying that idea might be for the missionaries. Only the wealthy few—that is, nations such as ours—have the power and the wherewithal to rescue us all from the impending consequences of mass consumption on a global scale. If we decline to do so, we will not be saved.

AT THE KYOTO CONFERENCE on climate change, in December 1997, the initial American position, egged on by the usual industrial interests, was that it would be unjust to impose target goals for reducing hydrocarbon pollution only on the advanced economies, since the emissions trend lines in developing countries are rising rapidly. But the U.S. gambit couldn't prevail on the world stage, and it didn't. More likely, the industry strategists were setting up a talking point for later use in American politics when seeking to sway public opinion and lobbying Congress to block any implementing legislation needed to achieve the goals.

A Brazilian diplomat at the Kyoto conference expressed the reaction of countries still trying to catch up with the wealthy: "They invite you in only for coffee after the dinner. Then they ask you to share the check, even though you didn't get to eat."

I am not arguing for irresponsibility among the striving poor of the world or global inattention to their wasteful practices. I am simply observing that the world will remain caught in a profound political stalemate on the environmental question until we Americans learn to put aside empty self-righteousness and accept the full burden of our historical position. As we already know, this is extremely difficult to do. It cuts against the popular sense of American triumphalism that is constantly promoted by the governing classes in politics as well as in business and finance. It makes us the center of the problem rather than virtuous spectators who are alarmed by what's happening.

In perverse ways, the global system may also help advance the process of self-knowledge because it allows us to see ourselves in stark relief. The action in the developing countries is like a loop of old film that continuously plays back our own history for us, making the distant facts of our past discomfitingly real. Nothing awful that is now happening in newly industrializing countries (even the practice of involuntary servitude) did not happen first, long ago, in the United States.

In early 1998, when the forests of Borneo were burning, sending sooty clouds across Southeast Asia, I was reminded of a mountain place in Vermont I have come to love and of the actual history of that verdant state. First, the indigenous people were driven out, robbed of their land, sometimes even killed. Then the giant white pines—the sequoias of New England—were swiftly harvested to provide ship masts until these great trees were virtually gone. Then old-growth forests were felled or burned to make mountainside pasture for sheep and cattle, followed by predictable, terrible flooding and erosion. When the New England wool industry collapsed a few decades later, people moved westward and, in different ways, repeated the process. They drained prairie wetlands to grow grain; they transformed the desert into a garden of cotton and artichokes.

The point is that it worked for us (albeit with a lot of breakage and suffering along the way). From the early stage of primitive capitalism, Americans accumulated the capital and incomes to build the predicate for higher development later on—a general prosperity based on industrialization and mass consumption. Serious people in the governing elite of developing countries know our true economic history, probably better than most Americans do, and they draw instruction from it. The environmental ethic, some of them conclude, is a veil of hypocrisy that conceals yet another version of old-fashioned neocolonialism.

They do have a point, but hypocrisy is not the important issue. The issue is industrial capitalism and its pathologies—its penchant for repeating, generation after generation across centuries, the same forms of abuse and exploitation previously thought to be extinguished and prohibited. As production and marketing are expanded to open up new territory, the shameful practices are revived on the frontiers of growth, and nobody stops them—the "dark Satanic mills" that William Blake decried, the easy returns derived from careless despoilation.

It took two centuries or more for Americans to develop the political power to abolish the most shameful human abuses. It took even longer for us to understand and then resist the industrial degradation of the natural world. Yet here we are with both again before us.

The poor nations struggling to be more like us (or at least less poor) cannot get at the systemic problem of capitalism any more than they can turn off the rising appetite for consumption. The latter drives the former, and indeed, the poor nations' burgeoning market demand (at least until financial crisis collapsed their growth rates) attracts our producers, too, eager to ride the same wave. This wheel keeps turning. Who really has the power to stop it? Aside from Ruskinian skeptics who wish for the pure and uncluttered life, who really wants to stop it? Think about this the next time you're on a 747.

I saw a living metaphor for the global dilemma on the streets of Beijing. The traffic flow at dusk along the Boulevard of Eternal Peace was an arresting spectacle—both graceful and disturbing— because China is on the edge of entering the automobile age that America embraced nearly a century ago. It's not quite there yet, but automobile ownership is increasing rapidly, and all the world's leading car makers have contested for a share of the market.

Beijing's traffic jams occur at boulevard intersections, where the cars and the bicycles meet, attempting to turn or to cross one another's paths. The swarms of bicyclists, gliding along the bike lanes like silent flocks of birds, are suddenly face to face with columns of cars and small trucks. Bumper to bumper, nobody yields; neither party backs off. The confrontation becomes a hopeless tangle of mismatched vehicles trying to inch past one another.

One may root for the bicycles, but it seems obvious that they are going to lose to the cars eventually, just as horses and pedestrians lost out generations ago in American cities. One may argue that China is mad to make this choice, that it should be patiently building railroads and urban mass transit systems instead. But Chinese planners understand that a strategy of patience is not the way to quickly acquire a world-class automobile industry that exports vehicles to the global market.

China has chosen cars, and so have the Chinese people who can afford them, for approximately the same reasons Americans love the automobile. These machines deliver real value to the human experience: speed and comfort, saved time and effort, the individuality of choice, and status.

The nightmare, of course, is the prospect of a China whose 1.2 billion citizens will someday be prosperous enough to consume automobiles at the same rate as do people in advanced countries. At present, China has 680 people per private automobile; the United States has 1.7 people per car. Could the world survive such progress? If not, who must give up their cars, the Chinese or the Americans? The answer seems obvious to the rest of the world.

Meanwhile, in America, the new consumer passion is owning an urban truck or a sport utility vehicle that conveys the get-out-of-my-way menace of a military vehicle. On which end of the global system does the madness lie?

The globalization of production has exposed the central fallacy that always lurked in the standard idea of industrial progress: one could believe in the notion that unending industrial expansion might eventually liberate everyone in the world from poverty only so long as there was no possibility that it would actually happen. Now that the world has been given a concrete glimpse of what that expansion would entail, the impossibility of expanding mass consumption becomes obvious. The finite limits of the earth collide with the human appetite for "stuff," and the impact leaves everyone gasping—people, plants, animals, the earth itself.

Yet the marketplace marches on. One response to the specter, an attitude that I suspect is widely felt if not always expressed, is a kind of high-minded environmental protectionism: shut it down. That is, stop the industrialization process before it kills us all. But I don't see that as a humane option or a very plausible one, for reasons of equity and politics.

From my observations, even people who are being shamefully abused by the emerging global system yearn for what it seems to promise them—the prospect of a wage income, an escape from perennial scarcity. Many indigenous peoples, certainly, are ensnared against their will (much as America's native people were swept aside by U.S. development). They need our help, for sure, but that is not the whole of the story.

It is a delusion, I think, for Americans to believe that the poorest people in the poorest countries do not really want industrial interference in their ancient state of muddy poverty. The great migrations that are under way around the world—the millions of people who leave hearth and home in a desperate search for work and wages—testify to the worldwide longing for a better life.

Despite their vast differences in culture and history, I believe that people everywhere, rich and poor alike, want the same elemental things in life: personal dignity and well-being, with some measure of control over their own destinies. In this, they are naturally drawn to the possibilities offered by electricity or motorized vehicles or indoor plumbing. To acknowledge the universal human yearning for material improvement does not excuse the patterns of destruction present in the global system or any of the cruelties it deals out to innocent peoples. On the contrary, the acceptance of our universality makes the random cruelties seem even more cruel, the destruction even more ominous.

Americans cannot escape responsibility for the global dilemma by blaming it on the underdeveloped sensibilities of citizens in the poorer countries or on the rapaciousness of some multinational corporations. We are their consumers, after all. Those new factories that generate new wealth in developing countries do so mainly by shipping shoes and shirts and toys, consumer electronics, semiconductor chips, steel, chemicals, even major components for cars and airplanes, to the wealthiest consumers of all.

For instance, the financial crisis that began in Thailand and spread across Southeast Asia is not really Asian but global. U.S. multinationals, banks, and financiers were full participants in constructing the bubble of overinvestment that collapsed, as were the Japanese and Europeans. In the same manner, it is no longer sufficient to identify the negative effects of development as Indonesian or Thai or even Chinese.

America exported its prosperity system, and the dynamics of its own history, as the model for others. It preached a doctrine of how to get "unpoor," aided and invested in the new players who followed the script, and occasionally punished some for their deviations. One does not need to tour those distant places to see that the global crisis of consumption is really America's, first and foremost. It is our model that's working for others, and it is not likely to change in fundamental ways until we show them how.

THE BRILLIANT POSSIBILITY of "one world" commerce that connects producers and consumers, workers and investors, across vast distances is the emerging recognition that there will be no place to hide. We will work out the terms for survival together or probably not at all.

The essential corollary is not so well understood, at least among Americans: economic inequality is, I think, fundamentally an environmental issue. I do not mean that everyone must become as rich in material goods as typical Americans or that rain forests should be paved over to make way for shopping malls. I mean, simply, that rising incomes and consumption, the process of industrialization itself, are necessary elements of any grand solution. This is true, obviously, among those peoples who are still confronted with perpetual scarcity, but it is also true *within* the wealthiest countries.

My point is about political reality more than morality. Any environmental action that simply pushes the costs down to lower rungs of the income ladder, whether the pain is imposed on impoverished nations or on working-class Americans, invites stalemate and the class-ridden political conflicts that are so easy for business interests to exploit. In the wealthiest countries, for instance, "green taxes" may produce that debilitating effect if the sponsors offer no offsetting relief for consumers at the low end of the food chain.

In poor countries, if development continues on more equitable terms, it should lead eventually to the leveling off of population growth rates, just as has occurred in the countries with advanced economies. To put it crudely, the surest way to promote middle-class behavior and public values is to ensure that people are able to achieve middle-class incomes. The manias of consumption that seem embedded in American life—more new toys for status, not comfort—may not necessarily appeal to other populations once they have established basic well-being and stable prosperity.

Even if that happy day should arrive, the larger dilemma must still be solved. The way out is not a secret. It involves nothing less than industrial transformation, both in production and consumption, a redefining of traditional ideas of economic growth in qualitative terms that eliminate rather than generate waste. As we know, technological processes already exist that could achieve much of this transformation, but most of them are only marginally applied.

What stands in the way more than the political power of the status quo interests are people's inherited attitudes—the existing expectations of consumers that are rewarded and reinforced in the marketplace. This is a formidable barrier, to be sure, but changeable. Because I still believe in democratic possibilities, I believe that as the public culture is altered, the industrial system can be made to follow.

People need a lot of help in learning how to think and behave differently. The work of reordering conventional thought has been under way for many years, and despite awesome resistance from entrenched interests, the struggle does actually make forward progress. The Kyoto conference on climate change, inadequate as the results may have been, was evidence of changing politics on a global scale. The next big breakthrough must be in changing economics.

One pioneering contribution in this regard is the work of Herman E. Daly, especially a book he wrote with theologian John B. Cobb Jr., *For the Common Good*.[1] I am among the many who have been educated by Daly's patient deconstruction (and demolition) of the scientific pretensions surrounding market economics. He is that rare economist wise and brave enough to stare own his own profession and describe the peculiar omissions and contradictions embedded in the economic model. In the standard model of production and consumption, the natural world does not exist and yet is presumed to be infinite. In real life, of course, the natural world is a finite storehouse of materials and a sink for all that is discarded and damaged.

Daly's insights into the true meaning of efficiency will change every calculation of profit and loss, progress and decline. Although most economists still resist his ideas they are the basis for a promising movement to redefine growth in qualitative rather than quantitative terms. This new framework can define a new economics in which growth once again becomes synonymous with genuine progress.

During my travels, another recurring delight was encountering people who are on the same page with Daly, even though they may speak a different language or approach the dilemma from other starting points. One of them was a Japanese industrial engineer named Hiroyuki Yoshikawa, who has worked on developing "social robots" to do jobs that are dirty or dangerous. When I met Yoshikawa, he was president of the University of Tokyo, the pinnacle of Japan's educational system. Instead of discussing robotics, he launched into a spirited explication of how the world can save itself.

"It is time for a new kind of revolution—a kind of humanized process of change that offers the only solution for our problems," Yoshikawa began. What exactly did he mean? He meant, literally, reinventing the industrial system, its processes and products, to complete the missing half. He envisions "a plus factory and a minus factory, a normal factory and an inverse factory"—a system that closes the input-output loop and protects nature even as it multiplies industrial employment.

"The process should be harmonized development . . . to improve the quality of life and also to develop this new kind of industry," Yoshikawa explained. "If we do this, if we develop this new dimension, we shall be free. We will invent a new industrial system and also solve our deepest social problems."

I came away exhilarated by Yoshikawa's confident optimism but also sobered by the difficulties of realizing his panoramic vision. In a sense, he was offering an engineer's version of Herman Daly's economics. The logic of both leads to radical change. But one cannot proceed very far down that road without, once again, bumping into the question of equity and economic inequality. The central goal, after all, is to unite the true costs of production with the market price of consumption. But how do people pay the price of higher quality if they are very poor or if their real incomes have declined while the minority enjoyed fabulous prosperity? Nor can society expect to coax or bludgeon private enterprise into accepting the full-cost pricing of goods if doing so would simply bring an active economy to a lurching halt.

Every environmentalist agrees that the cost-price principle is the goal, but I don't sense that much energy has been devoted to solving the underlying problems of income and inequality. That, too, requires radical change—a new social understanding that like the global system itself, we are all in this together now and one will not be saved unless all of us are.

In theory, these problems ought to be solvable if human attention and public spending are focused seriously on them. I can dimly imagine reforming the tax code and altering national priorities to create a system of negative and positive incentives in the marketplace or subsidy programs to develop the new production processes and new products that fulfill the visions of both Daly and Yoshikawa. Once the nation accepts that eliminating waste from everything is the central imperative, a multitude of targets pops up before us.

These are hard choices but not technologically implausible. Can we envision a universal car, available almost everywhere, that neither pollutes nor is discarded after a few years of use? Of course. The prototypes already exist. The question is whether ordinary folks will be able to afford them if they are produced. The government role is about making a market for the new (just as government made a market for armaments), but it is also about providing the financial aid many families need to purchase the same high-quality and durable goods that the upper classes insist on having.

In the end, these are political questions, not economic barriers, and there is no need to despair. If human ingenuity can invent the predicate for our destruction, then surely smart, good people can invent our way out of the dilemma.

What's Wrong with Consumer Society?

Competitive Spending and the "New Consumerism"

Juliet Schor

WHAT'S WRONG with consumer society? Although survey data suggest that there is strong public unease with the increasingly consumerist cast of American society, intellectual arguments against consumer society have failed to gain wide popular currency. Some founder on the shoals of elitism, paternalism, or a view of consumers as overly manipulable. Others, such as the antimaterialist messages of religion and morality and, to some extent, environmentalism, avoid these problems but fail to speak powerfully enough to the concerns of the average American. Why? One reason must surely be that the national consciousness remains enslaved to a liberal ideology that takes consumerism as unassailable.

Nowhere is that liberal ideology so powerful as in the discipline of economics. For economists, the answer to the question "What's wrong with consumer society?" is "Not a thing." Far from a problem, consumption is posited as a solution, ensuring well-being by eliminating pain and creating pleasure, or, in technical terms, giving "utility." Thus, consumption is the "good" that solves the problem of various "bads" (hunger, cold, boredom, etc.). For the most part, this approach emphasizes the functional or utilitarian characteristics of goods and services. Clothing keeps one warm or is aesthetically pleasing; food satisfies hunger or a discriminating palate; transportation moves one from place to place. Although such an emphasis is not dictated by the theory itself, personal prejudice and a political preference for conclusions that celebrate a free consumer market have led economists to an uncritical and simplistic approach to consumer behavior: virtually without question, whatever consumers do is in their own best interest. Such a hands-off stance has led the field of economics away from the social and symbolic functions of spending that are so prominent in anthropological, sociological, and literary analyses. Once we introduce these dimensions, however, economic analyses become at once more interesting and more critical. I begin with an essential difference in the way in which spending is viewed—as primarily an individual or a social act.

In the neoclassical view, the pattern of consumption is thought to emanate from a random distribution of individual tastes and preferences, as well as obvious variables such as family structure and income level. By contrast, in a social approach the distribution of tastes and preferences is not random across the population but corresponds to a definite structure, among whose defining characteristics are social and economic class. People of like class background have common tastes and consumption patterns. These similarities cannot be attributed only to functional needs (e.g., people with large families buy station wagons) but also are present in situations in which no or few functional considerations apply (e.g., taste in art or music, food, style of decor).

The fact that spending patterns vary by social class is nothing new. A hundred years ago, Thorstein Veblen, in his classic *Theory of the Leisure Class,* argued that "conspicuous consumption," that is, the visible display of discretionary spending, was the means by which individuals revealed their economic resources and thereby established social position.[1] Commodities "trickled down" a vertical class hierarchy in an emulative process occurring at each level. A recent, albeit somewhat more complex, analysis can be found in Pierre Bourdieu's *Distinction: A Social Critique of the Judgement of Taste.*[2] Bourdieu argues that not only economic class but also what he calls "cultural capital" affect consumption patterns. In his view, people acquire cultural capital through family socialization and educational background, and this cultural capital shapes their tastes and preferences. Taste becomes an expression of class position, as do the consumer choices associated with it. Bourdieu argues:

> Whereas the ideology of charisma regards taste in legitimate culture as a gift of nature, scientific observation shows that cultural needs are the product of upbringing and education: surveys establish that all cultural practices (museum visits, concert-going, reading, etc.), and preferences in literature, painting or music, are closely linked to educational level (measured by qualifications or length of schooling) and secondarily to social origin. . . . To the socially recognized hierarchy of the arts, and within each of them, of genres, schools or periods, corresponds a social hierarchy of the consumers. This predisposes tastes to function as markers of "class."[3]

What is the empirical evidence for the view that social class structures consumption? In the American, and to a lesser extent the British and Continental literature, earlier traditions that emphasized the class nature of consumption have fallen out of favor. U.S. research such as the classic "Yankee City" studies by W. Lloyd Warner and his colleagues has not been repeated in recent decades.[4] Surveys comparable to those on which Bourdieu relies for France do not exist for other countries. (There have, however, been some smaller-scale attempts, such as the work of Douglas Holt in the United States.)[5] What does exist is market research.

Private-sector firms, whose aim is to catalog and predict consumer expenditures and tastes, have developed extensive empirical models. Most are classificatory schemes (zip code, census block, or psychographic) used to predict spending patterns among various subsegments of any given population. Although these models have not been subjected to rigorous academic analysis, they are nevertheless useful. The most interesting among them are residential models, in which the census block—a smaller unit than the zip code—is found to be a strong predictor of household spending patterns. What residential classification schemes teach us is that consumption does remain structured by recognizable variables, which themselves correlate with various measures of social class. The patterns are not as clear-cut as they were sixty years ago, when one could easily decode class from the contents of a living room. There is today far more variation in patterns, as well as many more goods to account for, and there are also clear differences in *how* consumption occurs as well as just *what* is consumed. Nevertheless, an underlying social structuration still persists. Styles of furnishings, tastes in food, whether or not one watches public television, and preferences in clothing, cars, vacation destinations, and a wide range of other product choices remain differentiated by class and other social measures.

Spending patterns not only reflect a structure of social inequality but also reproduce it. Having proper taste, wearing the right clothes, and displaying certain manners are all means of achieving and then maintaining membership in a privileged social group. In Bourdieu's words, daily life is filled with "micro" acts of claiming status that lead to both inclusion in and exclusion from favored groups. The privileged use their habits of consumption to maintain their group identities and to exclude the less prestigious. This, for example, was the purpose of centuries of sumptuary laws that proscribed modes of dress and other spending activities. Socially visible, or "conspicuous," consumption is a major strategy used by high-status groups to keep themselves intact.

The role of spending in reproducing inequality is in some sense very modern. In previous eras, when status was determined by birth, history, and caste, spending played only a subsidiary role in the maintenance of social position. Consumption was constrained more by status position, as evidenced by sumptuary laws, cultural taboos not to spend "out of one's station," and so on. By the twentieth century in the United States and somewhat later in Europe, the consumption system had become far more open, and it was possible for a much wider range of individuals to spend as the rich or middle classes did (if they could find the income). Indeed, in societies in which birth, history, and caste are less prominent and status is a more fluid currency, consumption becomes more important. Urbanization, formal education, and the disappearance of traditional social relationships render spending more salient in establishing social position and personal identity. Thus, in the modern consumer society, commodities take on a new kind of symbolic importance. (Consumption has symbolic importance in all societies, but in consumer society its role in establishing personal identity and social position to some extent eclipse its symbolic role in ritual, religion, and so on.) More and more, what you wear and what you don't wear define who you are and where you are located on the social map. Although the social fluidity of the present is to be applauded, it exacts a price. Individuals face more pressure to use their income to gain access to a desired social group. This is particularly problematic in contexts in which failure to achieve middle-class status dooms one to a low "quality of life." In those cases, the pressures on individuals and households to spend in order to achieve some measure of status can be intense.

Thus, belonging to a particular social class now entails consuming a requisite set of goods and services. In such a world, there is always a dynamic process by which that requisite set of goods and services is upgraded, expanded, and modified. Within economics, this dynamic process is variously referred to as status, relative, positional, or, in my language, competitive consumption. In such approaches, the key feature is that consumption yields well-being or satisfaction not on the basis of its absolute level but always in relation to the level of consumption others have achieved. Those others comprise what sociologists term a reference group. Thus, when my neighbor acquires a new product, my own level of well-being falls, merely by virtue of my having fallen behind relatively. In order to avoid such a decline, I, too, must buy the new product, thereby "keeping up." Typically, new or upgraded products are adopted by a small group of innovating consumers, who initially increase their status by raising their relative position. Eventually, adoption of products becomes general as people attempt to reverse the decline in their well-being that has resulted from their failure to adopt. Thus, products diffuse throughout the population. Advertising and marketing that promote information about the products or about their growing prevalence can speed up the diffusion process, but diffusion would occur even without these efforts of producers.

In such accounts, the process by which information is conveyed becomes crucial. How do I learn about my neighbor's new acquisition? In small, open communities, such information is more or less transparent. We know one another well enough to visit one anothers' homes frequently enough to know what's being bought and by whom. In anonymous settings, the informational requirements are more complex and give rise to a situation in which competitive consumption occurs not with all goods but with a particular set of products. In order to figure in the status competition, the goods must be visible, or public, in their use and ownership. Clothing, housing, and automobiles have traditionally been such important status symbols because they are all accessible to public view and their use is easily verifiable. Savings, leisure time, insurance, and household furnishings and appliances that are not seen by visitors play a small role in the status-conferring process. This distinction between visible and nonvisible goods means that the former play a special and privileged role in the dynamic process. Because the competitive dimensions of spending are confined to this subset of goods, consumers often reduce their expenditures on nonstatus products in order to keep up with status goods. This occurs especially during periods when competitive spending is intensifying.

Classic postwar descriptions of the keeping-up process, such as those of James Duesenberry and Robert Frank, emphasize the role of proximate comparison, that is, comparison between individuals or families who are near each other in economic condition.[6] In particular, Duesenberry's influential account evoked a middle-class suburban world characterized by inclusion rather than exclusion. There were Smiths and Joneses, and they were very similar. In such a world, the middle class was growing, and popular wisdom had it that it would encompass all other classes. The nation and its spending patterns were homogenizing.

Beginning in the 1980s, those conditions changed, and what I have termed the new consumerism emerged. The new consumerism is more upscale in the sense that there is more aggressive, rather than defensive, consumption positioning. (It is similar to Veblen's account of the turn-of-the-century positioning among privileged groups.)[7] The new consumerism is more anonymous and is less socially benign than the old regime of keeping up with the Joneses. In part, this is because reference groups have become vertically elongated. People are now more likely to compare themselves with, or aspire to the lifestyles of, those far above them in the economic hierarchy. Microsoft's Bill Gates or a senior vice president have become more prevalent emulative targets.

A major reason for this change is the decline of the neighborhood as a prominent reference group. Because neighborhoods tend to contain individuals of similar incomes (houses are most families' major assets, and neighborhoods tend to have houses of similar value), using the neighbors as a standard kept people rooted in a proximate comparison. But as the neighborhood has declined as a focus of social interaction, so has its anchoring role. In its stead, the workplace arose as a fertile site for consumption comparisons. This process was accelerated by the growing numbers of married women who took on employment, particularly in white-collar and professional jobs. In the workplace, they were exposed to a more diverse reference group than was the typical suburban housewife and therefore were more likely to engage in upward consumption comparison (e.g., comparing themselves with superiors who had significantly higher incomes.) This perspective is supported by data from my own survey of roughly 800 employees of a major telecommunications company (hereafter the Telecom survey): only 2 percent of the sample identified their neighbors as their primary reference group, but 22 percent named their co-workers.[8]

As people spent less time in the homes of neighbors and even in the homes of friends, face-to-face socializing came to be replaced by television watching. Viewing hours have risen by about 50 percent since the mid-1960s and now are thought to occupy as much as 40 percent of adults' free time. Simultaneously, Americans of all classes have become increasingly concerned with privacy, shielding their homes from view by means of attached garages, fences, and decks. Thus, television has become increasingly important in providing information about the spending patterns of others. Television characters, our 1990s "friends," are a major source of consumption ideas, expectations, perceptions, aspirations, and comparisons. For example, in a 1991 survey, Susan Fournier and Michael Guiry found that 35 percent of their sample identified television commercials and magazine advertisements; and 27 percent identified television shows as a "really great idea source" for their own fantasy wish lists of things to get or buy.[9] In the words of consumer researchers Thomas O'Guinn and L. J. Shrum, "Television commonly uses consumption symbols as a means of visual shorthand; what television characters have and the activities in which they participate mark their social status with an economy of explanatory dialogue. Viewers see and hear what members of other social classes have and how they consume, even behind their closed doors."[10]

But television (as do other popular media such as films, advertisements, and lifestyle magazines) gives a heavily skewed picture of spending patterns, portraying almost exclusively the upper middle class and the rich. This leads to an inflation of Americans' perceptions of others' lifestyles. For example, O'Guinn and Shrum found that the more time people spend watching television, the more likely they are to believe that other Americans have tennis courts, private planes, convertibles, car telephones, maids, and swimming pools.[11] Heavy television watchers also have an exaggerated perception of the proportion of the population who are millionaires, have had cosmetic surgery, and belong to a private gym. Furthermore, the types of programs viewed also affect this upward distortion: soap operas (daytime or prime time) yield a larger upward bias than do other programs.

In my Telecom survey, I found a direct effect of television watching: it is correlated with spending more and saving less. Social theories of consumption hold that the inflation of norms raises aspirations, thereby leading to more spending. In my analyses, I found that every hour of television watched per week raised annual spending by $208 per year.[12] Another piece of evidence for the link between spending and television viewing is that debt and excessive television viewing appear to be correlated. In a poll conducted by the Merck Family Fund in 1995, the fraction responding that they "watch too much TV" rose steadily with level of indebtedness. More than half (56 percent) of those who reported themselves "heavily" in debt also said they "watch too much TV."[13]

As a result of television watching and new comparison processes, nearly everyone has begun observing and aspiring to the standards set by the upper middle class and the rich. The lifestyles of this group, which accounts for the top 20 percent of the income distribution, are approaching the status of cultural icons, looked to by those with far less income as increasingly necessary and worth having. Researchers Susan Fournier and Michael Guiry found that 35 percent of their sample of consumers aspired to reach the top 6 percent of the income distribution and another 49 percent aspired to the next 12 percent. Only 15 percent of their sample reported that they would be satisfied with "living a comfortable life," that is, being middle class.[14]

Another indicator of upscaling is that people are now more likely to believe that the good life can be had from material goods. Growing numbers of people believe that vacation homes, swimming pools, travel abroad, really nice clothes, a lot of money, and second cars are symbolic of a good life. Finally, the proportion of the population identifying various consumer items as necessities rather than luxuries has increased substantially since 1973.[15] The increasing prevalence and importance of brand-name status goods (as well as their cheap counterfeit versions) is another indicator of the growth of affluent lifestyles. Visible labels appear to have proliferated to a whole range of products that were not previously heavily "branded" or symbolized.

One reason why the top 20 percent has become so important as a lifestyle target is that this segment of the population's share of national income has increased dramatically. The shift began in the 1970s but accelerated in the 1980s and 1990s. And as the top 20 percent gained more—nearly half of the total income earned each year now goes to them—the 80 percent below earned less. Similarly, within the top 20 percent, the pattern has also become more unequal, with more income flowing to the top 5 percent. One consequence of this change has been an intensification of competitive spending. As a result of their increased income, the rich and super-rich began a bout of conspicuous luxury consumption, beginning in the early 1980s. Members of the upper middle class followed suit with their own imitative luxury spending. (Thus began the so-called decade of greed.) The 80 percent below, while gaining some ground in absolute terms, lost relatively to those above them. Not surprisingly, they emanated dissatisfaction and pessimism and engaged in a round of compensatory keeping-up consumption.

Thus, the competitive spending process has undergone a major, highly problematic change since approximately 1980. As members of the bottom 80 percent of the population have fallen behind relatively, they have become more inclined to imitate those in the top income group. The difference between what they aspire to and the income they have available to spend—what I call the "aspirational gap"—has increased enormously. As upscale lifestyles dominate aspirations, the aspirational gap grows. A majority of consumers find themselves structurally frustrated because their incomes are always inadequate to satisfy their desires. This dynamic has hit households in the $50,000–$75,000 income group particularly hard, contributing to the widely noted middle-class squeeze. (Not surprisingly, this is the group in which credit card debt has risen most dramatically.) Whereas in the days of proximate comparison the aspirational gap might have been on the order of 20 percent, it is now much higher. One survey of U.S. households found that the level of income needed to fulfill one's dream, that is, to satisfy aspirations, doubled between 1986 and 1994 and is currently more than twice the national median household income.[16]

One can speculate about the relationship between the aspirational gap and a range of dysfunctional consumer behaviors that have increased markedly since 1980. I refer here to a decline in household savings, a rise in credit card debt (especially among higher-income households), an increase in shoplifting, an increase in violent crime carried out to obtain particular status goods (athletic shoes, leather jackets, designer sunglasses), and the incidence of (and possible increase in) compulsive buying syndrome.

Indeed, it is tempting to speculate about a longer-term problem of consumer control. Taking the perspective of the entire twentieth century, one might ask whether human beings can adequately discipline themselves in our modern consumer paradise. On the one hand, traditional (or so-called primitive) constraints on ostentatious and luxurious spending, as well as religious and moral strictures on consumption, have eroded dramatically. On the other hand, the efforts of producers, advertisers, and marketers to create an alluring, even irresistible, spending environment have become ever more pervasive and sophisticated. What is the long-term effect of the new "religion" of consumerism that emerged nearly a hundred years ago, in which spending, and spending without limit, was extolled as something positive, therapeutic, and of benefit to the economy? One answer may be that we just cannot control ourselves in such an environment.

I have confined my discussion thus far to the ways in which consumption dynamics have changed in the United States. I believe, however, that these developments are also relevant in the new global economy. The growing influence of multinational corporations that distribute American consumer products, the rise of a worldwide popular media and electronic communications system, and worldwide trends in inequality suggest that the new consumerism may be spreading beyond the shores of the United States.

Perhaps most obvious is the increasing influence of U.S. con-

sumer product companies around the globe. They urge incorpo-
ration into a consumerist lifestyle: people are encouraged to give
up domestic, nonstatus versions of products; to switch from non-
commodified activities (such as teeth cleaning using a tree
branch) to commodified provision (toothbrushes and toothpaste);
or to acquire the new products offered by Western multinationals. This process is most highly developed in Europe, but it has
been growing substantially in Asia, Africa, and Latin America as
well, among both the middle classes and the poor. We are aware
of the most dramatic and scandalous of these examples: the association of infant mortality with formula feeding; the existence
of what has become known as commerciogenic malnutrition as
people substitute Coca-Cola and potato chips for healthier traditional foods; or the Avon ladies who paddle down the Amazon
River inducing poor women to spend large fractions of their meager incomes on cosmetics.[17]

But even apart from these dramatic examples, the long-term
role of branded Western products is worth considering. Although
it is true that branded products currently represent only a small
fraction of total consumption outside the industrialized countries,
they are central to the operation of a competitive consumption
model, and their growth is laying the groundwork for its proliferation and deepening. Furthermore, other behavioral aspects of
the American companies are worth considering. These include
shortening of the product life cycle; high levels of advertising and
marketing relative to production costs (i.e., a high symbolic content to goods); an emphasis on what has been called commodity
aesthetics (i.e., high investment in the aesthetics of design); and
ecological unsustainability in production and use.

Finally, as American and other Western popular media become
more important around the globe, we can expect them to play an
increasing role in setting consumer aspirations. The 20 percent
group from the West may increasingly become the worldwide aspirational standard. Just as Americans who are heavy television
watchers come to believe that a swimming pool or a luxury car is
an American consumption norm, so, too, will villagers in China
and Brazil. An affluent, out-of-reach lifestyle will increasingly

seem normal and hence necessary to attain. A profound aspirational gap has already appeared and may well grow. That gap will exacerbate pressures from elite and middle-class groups to increase their share of national income.

Thus, on a global basis, consumer culture may well intensify a competitive spending process in which there are few limits, in which the aspirational gap is ubiquitous and growing, and in which alternatives that have been shown to contribute far more to human well-being (leisure, savings, public goods) are crowded out by private status goods. This would constitute a profound failure at the heart of the global economy.

The foregoing analysis suggests a number of arguments for constructing a more rigorous, persuasive, and comprehensive critique of consumer society than is often found in the critical literature. Three arguments present themselves. First, there is a self-defeating aspect to competitive consumption. Where what matters are relative rather than absolute levels of consumption, general increases in spending do not raise utility but leave people in the same position as before the increase. In the extreme case, where all utility is positional, general spending increases confer no additional utility at all. This is the so-called prisoners' dilemma aspect of the model—everyone would be better off cooperating because consumption has costs such as labor expended, natural resources used up, and so on. But without an entity to create cooperation, a bad outcome for everyone results. The extent to which this prisoners' dilemma scenario actually characterizes current consumption is, of course, an empirical question, but the evidence on income and happiness suggests that general increases in income do not yield improvements in self-reported happiness and well-being. The evidence is very consistent with a self-defeating system, a treadmill, in effect.

A second problem with competitive consumption is that the pressure to keep up in acquiring visible, private status goods crowds out other, competing uses of income. The four major competing uses of income are leisure, savings, public goods (including the environment), and nonvisible private consumption. The experience of the past two decades in the United States suggests the plausibility of such a dynamic. Working hours have risen substantially; the 1997 household savings rate (3.9 percent) was the lowest in sixty years; public expenditures have been reduced dramatically in order to reduce taxes and the public deficit. If quality of life is produced by a variety of uses of economic resources including free time, high-quality public goods, and financial security, the intensification of pressures to spend on status items produces an unfortunate competitive outcome.

Finally, the rise of an aspirational gap creates a persistent dissatisfaction among consumers that cannot be cured at any level of absolute income. If what people want is determined largely by what an affluent group with rising incomes has, large numbers of people will be left with the belief that they have not achieved enough. This yearning, along with the sometimes destructive behaviors associated with it, creates an ongoing tragedy of modern consumer society.

Consuming for Love

Edward N. Luttwak

AMERICANS PROTEST their superior love of individual freedom, with much historical justification. Yet they enslave themselves to demon debt just to accumulate all sorts of far-from-essentials, from large, powerful trucks used as mere cars down to porcelain baubles advertised on late-night television ("Valuable instant heirlooms for only five easy payments of $19.99!"). To pay for their buying habit, Americans work more hours during each year than do members of any other advanced population on earth except for the Japanese. When it comes to vacations, the Japanese again are ahead, at twenty-five days per year as against the twenty-three that Americans average—a miserable portion of free time compared with the Germans' forty-two days and the thirty-eight days that the French find insufficient.

True, some people obtain so much satisfaction from their jobs that they live to work. But many of those who work only for money are eager to work overtime and even seek second jobs, sacrificing their personal freedom and family life just to consume more. Actually, many Americans do not choose to work in order to buy—they *must* work to pay interest and repay principal on what they have already bought.

Frugal East Asians save a substantial portion of what they earn, setting aside as much as one-half of their personal incomes in hard-striving China and one-third of far larger incomes in Japan. Prudent Europeans save roughly one-quarter. Americans, by contrast, save very little, and the amount is becoming less and less— of late not quite one-twentieth of their personal incomes. Even that phenomenally low proportion represents an average that is heavily skewed by the large savings of the highest-earning households. Most Americans, in fact, save less than nothing, borrowing with abandon from all possible sources: from credit card issuers at very high interest, from home equity lenders at the risk of losing their dwellings, from banks and credit unions up to their credit limits, from mortgage lenders for the largest sums, and from pawnshops for the smallest.

Americans scarcely invented debt, but three things are unique about their indebtedness. The first remarkable characteristic is its inordinate dimensions, which keep increasing. By mid-1997, the total debt of all American households reached the unprecedented level of 89 percent of total household income. It is no coincidence that the foreign debt of the United States is now by far the largest ever recorded for any country at any time in history, for domestic savings are so small compared with both household and government debt. By contrast, Italy's notoriously huge government debt, some 120 percent of the country's gross national product, is offset by equally huge domestic savings, such that Italy is actually a net lender to the rest of the world.

The second unique characteristic of American indebtedness is its dissociation from poverty, the traditional cause of personal and family debt. Indeed, the poorest 20 percent of American households owe very little to anyone, for nobody except petty money-lenders with sturdy enforcers allows them any credit. The recent increase in so-called subprime lending, which carries especially high interest rates and is much employed in the selling of used automobiles, largely reflects the borrowing of the next-to-poorest 20 percent of American households—which are far from poor by any historical or international standard. Most American borrowers are not poor at all, or, rather, they would not be poor but for their borrowing to spend more than they earn.

The third characteristic is the particular use that is made of the immense amounts borrowed. Indian peasants go into debt to feed their families when the monsoon fails and to marry off their daughters; young couples all over the world borrow to buy their first household necessities. A large portion of American household debt likewise consists of home mortgages and college loans, but much of it reflects the purchase of expensive motor vehicles, designer clothing, signature watches, assorted recreational gadgets, and all manner of other things that are scarcely necessities of life by anyone's definition. To borrow at 18 percent or more on a credit card to purchase a fancy bit of clothing is a commonplace of American life; to exhaust the equity limit in a second mortgage to buy a luxury car, leaving no margin of safety from dispossession, is by no means unusual—even dealers of Mercedes-Benz and BMW automobiles encounter few cash buyers.

Fearing the ultimate consequences of savings so small and spending so great that the U.S. foreign debt is passing $1 trillion and is on its way to twice that level, even rigorously values-free academic economists have lately begun to insert sotto voce pleas for a bit of frugality. Others more boldly ask for a restoration of all forms of Calvinist restraint, Calvinist compulsion, and Calvinist punishment, with enormous success in every sphere but one. Antipornography, antismoking, antifat, anti–beach nudity, anti-sugar, antisex, antinarcotic, and antialcohol campaigns are vigorously advancing while more prison terms, longer prison terms, mandatory life sentences, dozens of new death penalty statutes, accelerated executions, ever harsher prison conditions, and even a return of chain gangs all reveal how the economic insecurities of today's "turbo-charged capitalism" are vented.

In the past, this would have been accomplished more simply by the harsh persecution of a targeted minority group, preferably racially distinct. It alone would have paid the price for the declining relative incomes of many Americans and for the lost job security of many more in the brave new economy of unlimited competition and endless structural change. This new economy offers splendid opportunities to financial acrobats but keeps many other Americans awake at night, fearfully wondering what the morrow will bring.

Today, however, with the persecution of minorities ruled out both by fashion and by many laws, all those unexpressed fears and all the anger of insecure breadwinners can find an outlet only in the prohibition of all that can be prohibited, including nude pixels on the Internet, and in harsh punishment by way of a suitably named criminal justice system, which now holds 1.8 million Americans behind bars.

Only one form of Calvinist restraint has not advanced at all. A solid wall of conventional opinion (with President Bill Clinton running just ahead of the crowd) loudly supports all the "anti" campaigns, but not even a fringe group has tried to condemn the borrow-and-buy habit that is by far the broadest American addiction. Thus, the most important of all the original Calvinist virtues, saving—capital accumulation and investment in lieu of consumption—is the only one that remains forgotten. The proclivities of the professional evangelists who lead many of the "anti" campaigns are no doubt a factor—people so eager to buy beach condominiums, jewelry, and costly cars with the contributions of their devotees are justifiably hesitant to deplore any form of materialism. Also a factor is the deeply American conjunction between moralism and greed, a perverted rendition of the original Calvinist belief that virtue is rewarded by wealth.

In fairness to crassly materialistic evangelists and greedy moralists, however, it must be recognized that there is a sound justification for their failure to include the consumer addiction in their sweeping condemnations of every other form of self-indulgence. For there is nothing frivolous about the buying habit. It may be self-indulgence, but its motives derive from the most profound of human needs. *Homo americanus* is genetically programmed to live with the constant emotional support of an entire family, just as much as is the next *Homo sapiens* down the road in Norway or Italy, but most often lives in a state of emotional solitude to which the species has not yet adapted.

Leopards do all right in habitual solitude, except when they are mothers rearing cubs before their very rapid weaning. Hyenas and baboons, on the other hand, are not self-sufficient individuals but components of extended families or clans in which they are variously petted, comforted, protected, reassured, disciplined, and obediently followed at different stages of life by an entire multigenerational array of blood relatives. There are, of course, hyenas and baboons that for one reason or another are separated from their clans; they either die quickly or survive only as frantic rogues or melancholy neurotics. As it happens, in its present state of evolution, *Homo sapiens* is just like the hyena and the baboon. Yet most Americans must live without the emotional support that their genes require.

In normal human societies, in which extended-family ties are naturally preserved by geographic proximity from birth till death or are actively sustained by those who depart to live elsewhere, the annual calendar is a sequence of birth celebrations, religious or voluptuary festivals, marriages, anniversaries, and funerals, all of which also serve as clan reunions. With much hugging and kissing of the very young and tacit pledging of mutual aid and comfort among the less young, family ties are maintained, repaired, and strengthened. Uncles and aunts are reaffirmed as multiple parents at one remove, just as the cousinage of their children is an assemblage of siblings at one remove, granduncles and grandaunts are supplementary grandparents, and even second cousins still function as blood relatives in working order. From all these, material aid and emotional support are expected, received, and reciprocated.

As against this—that is, normal human behavior—most Ameri-

cans are emotional destitutes, as poor in their family connections
as are Afghans or Sudanese in money.

Many Americans still marry, of course, but today's marriages,
now presumptively fragile even if unbroken, supply much anxiety
as well as support, failing to serve as even fragmentary replace-
ments for an entire family in working order.

For American children, uncles and aunts are in most cases only
distant presences. They are sources of infrequent small gifts at
best, objects of occasional humorous anecdotes ("We once visited
Uncle Charlie, who lives in California, and although it was a very
warm day . . .") or sinister legends ("I heard that Uncle Bill—who
I never met because he lived in Maine—was accused of . . .")
rather than sources of quasi-parental caring and support. As for
second cousins, they are little more than strangers if they are
known at all, and cousins rarely accept any mutual obligation.
Even siblings strictly limit their responsibilities to one another—
not a few of the homeless people found dead after a cold winter's
night have brothers or sisters who live in some comfort and who
discharged all fraternal responsibilities long ago, perhaps by a few
dollars grudgingly given out on a doorstep.

American parents have their children, of course, but to hug is
not the same thing as to be hugged. Besides, the same societal
conditions that leave parents in solitude induce children to live
apart from them—indeed, the youth living at home beyond his
teens is considered not a caring child but a problem, even an embarrassment.

True, this predicament is probably only temporary. In another
20,000 years or so, *Homo sapiens* may well adapt to the present
condition of most American lives. In the meantime, with much
ingenuity, the vast majority of affected Americans have found
their diverse ways of evading the suicidal aggression or deadly
melancholy of isolated hyenas and baboons. First, they fully participate in the abandonment of family ties by way of christenings,
weddings, and funerals forgone in favor of ballgames and fishing
and golf; by way of departures to distant places for even fractional
economic advancement; by way of theme park or resort vacations
in lieu of a round of familial visits; by way of telephone calls un-

made, letters unwritten, and gifts ungiven. Having done this, they find a substitute, which may be only processed cheese rather than Brie or Stilton, but which certainly has the sovereign American virtues of cheapness and immediate availability.

This emotional equivalent of fast food is supplied by the chapels, temples, ashrams, sanctuaries, and covens of pseudoevangelical, New Age, pseudo-Hindu, pseudo-Christian, pseudo-Buddhist, pseudo-Muslim, Catholic, pseudoscientific, pseudopolitical, and many other cults. The United States is filled with them. Each offers its modestly or expansively charismatic imitation of a father who instructs his children with some admixture of severity and paternal love. Each strives to provide an imitation family. Often, there is much encouragement of hugging, kissing, or at least hand-holding by the devotees in attendance, though some cults go in for the severe style, stressing strict obedience of exacting rules, and a few offer only ideological intensity. Most of the cults specifically promote themselves as fostering reciprocal caring, even love, in contrast to the established religions' focus on obedience to some god or other. Almost all emphasize the emotional warmth of their assemblies, as opposed to the cold ceremonialism of established churches.

The combined revenues of today's roster of American cults probably exceed those of the computer industry—some reach $300 million per year, and many are in the tens of millions. Going by the only criterion that matters to most of their leader/owners, the cults are certainly a great success.

But they are also cost-effective for their devotees in many cases. Money aside—though some cults demand a great deal of it—the only cost is to the intellect, as devotees must willfully suspend their critical faculties in order to believe implausible and sometimes exceedingly bizarre doctrines. That sacrifice, however, usually amounts to very much less in terms of time, attention, and even money than the years of hugging, kissing, telephoning, writing, giving, traveling, visiting, listening, waiting, and attending the sick that the upkeep of natural families requires. Instead of all that, most cultists need only drive a short way to their chosen franchise or corporate headquarters, park their car, and presto!

They can plug into a simulated instant family, often complete
with earnest expressions of loving concern from fellow devotees
within range.

So much success in attracting the lonely crowd could not be ig-
nored by the established churches. Many have reacted as smart
competitors always will. They, too, now favor hand-holding and
hugging. There is with this a more or less subtle shift in emphasis,
from the delivery of obedience and love to the deity to the supply
of love to the congregants themselves, provided by the congregation as well as by the deity conscripted to the task. Emotional fast
food can now be had from millennial or at least secular churches
as well as from myriad cults much less antique than the International Business Machines Corporation.

Other Americans cope in more perilous ways, by substituting a
private universe of habitual, frequent, or occasional alcoholic or
narcotic stupefaction for the missing emotional security of a sup-
portive clan. This replacement, too, is mostly successful, contrary
to legend. Instead of embarking on a brief career as rogue hyenas
or mad baboons, destined to be killed in their first incautious assault, instead of willing themselves to die, as clan animals locked
in cages often do, emotionally isolated Americans drink, inhale, or
inject, as the case may be, while effectively pursuing their careers
and maintaining their fragment of a family—the very same nuclear families of nonaddicted Americans. As for the monetary
cost, it is often trivial. Even the fanciest illegal substances hardly
make a dent in the budget of many users, some of whom are multi-
millionaires, many more of whom are merely affluent. The ruined
drug addict on his way to a penniless death in a sordid alley is
mostly myth. If this were not so, if the customer base truly con-
sisted of derelicts and the unemployable as well as petty criminals, Colombian drug lords would have to be content with tene-
ments and bicycles instead of palatial mansions and executive
jets.

The trouble with cults and psychoactive chemicals is that both
provide only intermittent relief from the chronic emotional deficit
of a clan species in unnatural solitude. Besides, many Americans

refuse both options for reasons of temperament, intellect, or prudence.

Not all chemical remedies are disruptive or illegal. According to the latest statistics on obesity, one-half of all Americans find their emotional fast food *in* fast food or, rather, in all foods—fast, slow, instant, snack, or sweet. Americans are collectively envied in much of the world for their political, economic, and military successes, but their silhouettes alone show that many of them (excluding those whose obesity is caused by endocrine dysfunction) consider themselves to be hopeless failures.

And there is, of course, work and its satisfactions, which can certainly overcome almost any emotional deficit or even recycle it into a reason to do more work. Conventionally viewed with more approval than cults, chemicals, or overeating, overwork is a peculiar remedy for family ties in disrepair because it is itself the greatest destroyer of family ties. That makes it no less effective as a substitute—but only for the minority whose work is persistently exciting or deeply purposeful, from racing-car drivers to scientists gripped by their research, from tycoons on a roll to genuinely devoted hospice nurses.

Each of the foregoing remedies is effective up to a point, but by far the most common and most important of remedies is morale boosting by a constant stream of presents.

It is always very nice to receive presents, but the presents in question here are not gifts; they are not given by others. Because the predicament itself implies a sad insufficiency of parental, filial, avuncular, or cousinly loving care, *Americans instead buy presents for themselves.* And this they do—in shops and department stores; by responding to advertisements, mass-mailed catalogs, and direct-mail and telephone solicitations; by telephoning in to order objects displayed on television shopping programs; and, lately, by ordering on-line through the Internet—on a colossal, multitrillion-dollar scale.

Because an extraordinary proportion of what is bought by Americans consists of presents, including some gifts to others, the essential character of the consumer goods on offer has become steadily less functional. Much of what Americans buy nowadays amounts to baroque elaborations in which the object's original functional purposes are submerged by excesses, fripperies, frills, or simply overdesign. There are entire categories of once practical products, from baggage to writing implements, that now appear in the guise of "executive toys." Any puritanical outrage at other people's enjoyment of fancy goods is really beside the point: the peculiar preferences of American buyers have enormous economic consequences because they cripple U.S. exports of consumer goods. Buyers around the world who are not in need of boosting their morale refuse to buy "gift" versions of the objects they seek. Most notably, today the U.S. automobile industry produces very few unadorned vehicles meant to provide a means of transportation. Instead, it turns out heavyweight luxury passenger "trucks," enormous "family vans," coupés of deliberately reduced functionality (they cost more yet provide less room), convertibles, and sports cars, as well as a mass of super-lavish and intermediately lavish sedans.

Great fun is had by all, but automobile buyers in other countries reject the results for reasons of fuel economy, plain economy, environmental economy, or just good taste. The result is manifest in U.S. trade figures. Instead of contributing to a favorable trade balance, the automobile industry, by far the largest of all U.S. manufacturing industries, is so weak an exporter that the sector accounts for the single biggest item in persistent U.S. trade deficits. According to the latest figures as of this writing, between January and July 1997, the United States imported motor vehicles for a total of $65.3 billion, as opposed to $31.8 billion in exports, resulting in a deficit of $33.5 billion, almost one-third of the overall trade deficit of $108.4 billion during that period.

The same is true of other categories of consumer goods, so much so that the United States is now an exporter primarily of capital equipment, notably passenger aircraft, precision machinery, and power-generation equipment, as well as chemicals—all sectors that are uncontaminated by the baroque propensities of American consumers out to buy presents for themselves.

The larger economic picture reveals larger consequences. In

the booming U.S. economy of 1996, whose gross domestic product reached $6.9 trillion, personal consumption expenditures for goods and an increasing proportion of services reached $4.7 trillion—an extraordinarily high proportion (68 percent) by world standards.

The consumption of some services is, of course, entirely involuntary, as in the case of emergency, chronic, and terminal medical care as well as the legal services all too many Americans must purchase to defend themselves in lawsuits. None of those expenditures reflects a hidden emotional agenda. At the opposite extreme are the highly personal services bought largely as self-given presents (hair care, massage, manicure). In between these two limiting cases, there are two interesting categories of consumption expenditure: services for the elderly and tourist travel.

Americans spend more and more for old-age homes, assisted-living residences, and similar euphemistic variants of the same. Conventionally, the increasing consumption of such services is readily explained as a straightforward demographic phenomenon. That it is, though greatly modified by the consequences of family breakdown—except that the families in question consider themselves to be perfectly normal and unbroken, even admirable, models of family life. Matters have evolved so far that the phrase *family breakdown* invariably refers to underclass nonfamilies, wherein women without husbands are left to care for children and grandchildren, with public assistance sparingly given and lately often denied. On the other hand, affluent couples who live in spacious suburban homes and yet consign their parents to the care of strangers are nevertheless not considered broken families. That many elderly parents may actually prefer to live apart from their children, and out of reach of their grandchildren except for more or less occasional visits, indicates nothing except the degree to which family breakdown has been internalized as normal. One of the perversities of national accounting, incidentally, is that the U.S. gross national product is all the greater for the fact that care of the elderly devolves into commercial transactions that generate recorded sales. In contrast, the gross national product of Spain, for example, is diminished because grandparents live with their children and grandchildren, generating no numbers that statisticians can collect but providing the firmest possible basis for the emotional security of all involved.

As for tourist travel expenditures, Americans are notably modest compared with Europeans, who in addition to doing a great deal of family visiting are far more inclined to travel to exotic destinations at any given level of disposable income. This is an interesting exception because tourist travel is rarely individual; it is most unusual for couples not to travel together, with or without children. One consequence is that this category of expenditure does not afford much opportunity for self-given presents meant to assuage feelings of emotional isolation.

In any case, what is squeezed out by so much personal consumption of goods and services is government expenditure and investment ($1.2 trillion in 1996) and private investment ($1.0 trillion in 1996, a boom year withal). This is in spite of the modest but useful contribution of the outside world by way of a $144 billion trade deficit in goods and services—on credit, of course—so that during the year 1996, the United States once again added to its total foreign debt by overconsuming.

The low level of government expenditure and investment yields the paradox of a very rich country with federal, state, and local levels of government so poor that they cannot afford to provide health care to citizens as every other developed country does in one way or another or to assist the poor as they are assisted in every other affluent country.

As for the low level of private investment, it at least contributes to the much debated and much deplored lag in labor productivity, that is, the very slow (1 percent per annum) rate of increase in the total product obtainable from given amounts of labor.

The great claim that is made for the flexibility of Americans and hence of the U.S. labor market—that is, the willingness of Americans to move from place to place, to change their trade or even their profession, and to accept lower wages in order to keep working—is, of course, its economic efficiency. Nobody pretends that the personal and social consequences of so much mobility and adaptability are positively desirable in themselves.

However, when the chain of repercussions is followed step by step, from an economic system that achieves efficiency by imposing constant structural change, to the fragmentation of family life that is thereby caused, to the resulting psychological consequences, to the consumption habits these consequences induce, and then to the effect of those habits on the economic system by way of low savings and baroque consumer preferences, it may be concluded that a more rigid but more stable economy could be even more efficient. Labor costs would be higher, but so would savings rates because less mobile, more secure employees would be more likely to plan ahead and to sustain reassuring family ties, reducing their emotionally motivated consumption expenditures. A higher savings rate would increase the supply of capital, which would in turn increase the productivity of labor, offsetting higher labor costs.

At a time when the dynamic "turbo-capitalism" of today's deregulated, globalized U.S. economy is greatly celebrated, one may recall that until the late 1970s, the U.S. economy *was* more rigid and much more stable. This was because a great many industries, from airlines to natural gas companies to savings and loan associations, were subject to detailed regulation and thus were stabilized, as were their labor forces. At the time, needless to say, consumption spending of all kinds absorbed a lower proportion of incomes. Thus savings rates were higher—and so was growth, for all the supposed inefficiencies of regulation. There were fewer opportunities for financial acrobats to accumulate enormous wealth (the chief executive officers of regulated airlines earned the miserable pittance of only $1 million per year), and employees at large were economically secure because stable industries meant stable jobs. Judging by the lower voluptuary spending of the period, Americans were also happier.

False Connections

Alex Kotlowitz

A DRIVE DOWN Chicago's Madison Street, moving west from the lake, is a short lesson in America's fault lines of race and class. The first mile runs through the city's downtown—or the Loop, as it's called locally—past high-rises that house banks and law firms, advertising agencies and investment funds. The second mile, once lined by flophouses and greasy diners, has hitched onto its neighbor to the east, becoming a mecca for artists and new, hip restaurants, a more affordable appendage to the Loop. And west from there, past the United Center, home to the Chicago Bulls, the boulevard descends into the abyssal lows of neighborhoods where work has disappeared. Buildings lean like punch-drunk boxers. Makers of plywood do big business here, patching those same buildings' open wounds. At dusk, the gangs claim ownership to the corners and hawk their wares, whatever is the craze of the moment, crack or smack or reefer. It's all for sale. Along one stretch, young women, their long, bare legs shimmering under the lamplight, smile and beckon and mumble short, pithy descriptions of the pleasures they promise to deliver.

Such is urban decay. Such are the remains of the seemingly intractable, distinctly American version of poverty, a poverty not only "of the pocket" but also, as Mother Teresa said when she visited this section of the city, "of the spirit."

What is most striking about this drive down Madison, though, is that so few whites make it. Chicago's West Side, like other central-city neighborhoods, sits apart from everything and everyone else. Its inhabitants have become geographically and spiritually isolated from all that surrounds them, islands unto themselves. Even the violence—which, myth has it, threatens us all—is contained within its borders. Drug dealers shoot drug dealers. Gang members maul gang members. And the innocents, the passersby who get caught in the crossfire, are their neighbors and friends. It was that isolation which so struck me when I first began to spend time at the Henry Horner Homes, a Chicago public housing complex that sits along that Madison Street corridor. Lafeyette and Pharoah, the two boys I wrote about in my book *There Are No Children Here,* had never been to the Loop, one mile away. They'd never walked the halls of the Art Institute of Chicago or felt the spray from the Buckingham Fountain. They'd never ogled the sharks at the John G. Shedd Aquarium or stood in the shadow of the stuffed pachyderms at the Field Museum. They'd never been to the suburbs. They'd never been to the countryside. In fact, until we stayed at a hotel one summer on their first fishing trip, they'd never felt the steady stream of a shower. (Henry Horner's apartments had only bathtubs.) At one point, the boys, so certain that their way of life was the only way of life, insisted that my neighborhood, a gentrified community on the city's

North Side, had to be controlled by gangs. They knew nothing different.

And yet children like Lafeyette and Pharoah do have a connection to the American mainstream: it is as consumers that inner-city children, otherwise so disconnected from the world around them, identify themselves not as ghetto kids or project kids but as Americans or just plain kids. And they are as much consumers as they are the consumed; that is, they mimic white America while white America mimics them. "Inner-city kids will embrace a fashion item as their own that shows they have a connection, and then you'll see the prep school kids reinvent it, trying to look hip-hop," says Sarah Young, a consultant to businesses interested in tapping the urban market. "It's a cycle."[1] A friend, a black nineteen-year-old from the city's West Side, suggests that this dynamic occurs because the inner-city poor equate classiness with suburban whites while those same suburban whites equate hipness with the inner-city poor. If he's right, it suggests that commercialism may be our most powerful link, one that in the end only accentuates and prolongs the myths we have built up about each other.

Along Madison Street, halfway between the Loop and the city's boundary, sits an old, worn-out shopping strip containing small, transient stores. They open and close almost seasonally—balloons mark the openings; "Close-Out Sale" banners mark the closings—as the African American and Middle Eastern immigrant owners ride the ebb and flow of unpredictable fashion tastes. GQ Sports. Dress to Impress. Best Fit. Chic Classics. Dream Team. On weekend afternoons, the makeshift mall is thronged with customers blithely unaware that store ownership and names may have changed since their last visit. Young mothers guiding their children by the shoulders and older women seeking a specific purchase pick their way through packs of teenagers who laugh and clown, pulling and pushing one another into the stores. Their whimsical tastes are the subject of intense curiosity, longing, and marketing surveys on the part of store owners and corporate planners.

On a recent spring afternoon, as I made my way down Madison Street toward Tops and Bottoms, one of the area's more popular shops, I detected the unmistakable sweet odor of marijuana. Along the building's side, two teenage boys toked away at cigar-sized joints, called blunts. The store is long and narrow; its walls are lined with shoes and caps and its center is packed with shirts, jeans, and leather jackets. The owner of the store, a Palestinian immigrant, recognized me from my previous visits there with Lafeyette and Pharoah. "You're a probation officer, right?" he asked. I told him what my connection was with the boys. He completed a sale of a black Starter skullcap and then beckoned me toward the back of the store, where we could talk without distraction.

Behind him, an array of nearly 200 assorted shoes and sneakers lined the wall from floor to ceiling. There were the predictable brands: Nike, Fila, and Reebok, the shoes that have come to define (and nearly bankrupt) a generation. There were the heavy boots by Timberland and Lugz that have become popular among urban teens. But it was the arrangement of shoes directly in front of me that the proprietor pointed to, a collection of Hush Puppies. "See that?" he asked. "It's totally white-bread." Indeed, Hush Puppies, once of earth tones and worn by preppies, have caught on among urban black teens—and the company has responded in kind, producing the shoes in outrageous, gotta-look-at-me colors such as crayon orange and fire-engine red. I remember the first time Pharoah appeared in a pair of lime green Hush Puppies loafers—I was dumbfounded. But then I thought of his other passions: Tommy Hilfiger shirts, Coach wallets, Guess? jeans. They were the fashions of the economically well heeled, templates of those who had "made it." Pharoah, who is now off at college, ultimately found his path. But for those who are left behind, these fashions are their "in." They give them cachet. They link them to a more secure, more prosperous world, a world in which they have not been able to participate—except as consumers.

Sarah Young, whose clients include the company that manufactures Hush Puppies, suggests that "for a lot of these kids, what they wear is who they are because that's all they have to connect them to the rest of the larger community. It marks their status because there's not a lot else."[2]

It's a false status, of course. They hold on to the idea that to "make it" means to consume at will, to buy a $100 Coach wallet or an $80 Tommy Hilfiger shirt. And these brand-name companies, knowing they have a good thing going, capitalize on their popularity among the urban poor, a group that despite its economic difficulties represents a surprisingly lucrative market. The companies gear their advertising to this market segment. People such as Sarah Young nurture relationships with rap artists, who they lure into wearing certain clothing items. When the company that makes Hush Puppies was looking to increase their presence in the urban market, Young helped persuade Wyclef Jean, a singer with the Fugees, to wear powder blue Bridgeport chukkas, which bear a sneaking resemblance to the Wallabee shoes familiar to members of my generation. In a recent issue of *Vibe*, a magazine aimed at the hip-hop market, rappers Beenie Man and Bounty Killer are pictured posing in Ralph Lauren hats and Armani sweaters, sandwiched between photographs of other rappers decked out in Calvin Klein sunglasses and Kenneth Cole shoes. The first three full-page advertisements in that same issue are for Hilfiger's athletic line, Coach handbags (with jazz singer Cassandra Wilson joyfully walking along with her Coach bag slung over her shoulder), and Perry Ellis casual wear (with a black man and three young boys lounging on the beach). This, as Pharoah told me, represents class—and, as Young suggested, the one connection that children growing up amid the ruins of the inner city have to a more prosperous, more secure world. It is as consumers that they claim citizenship. And yet that Coach handbag or that Tommy Hilfiger or Perry Ellis shirt changes nothing of the cruel realities of growing up poor and black. It reminds me of the murals painted on abandoned buildings in the South Bronx: pictures of flowers, window shades, and curtains, and the interiors of tidy rooms. As Jonathan Kozol observes in his book *Amazing Grace,* "the pictures have been done so well that when you look, the first time, you imagine that you're seeing into people's homes—pleasant-looking homes, in fact, that have a distinctly middle-class appearance."[3]

But the urban poor are more than just consumers. They help drive fashions as well. The Tommy Hilfiger clothing line, aimed initially at preppies, became hot in the inner city, pushed in large part by rap artists who took to the clothing maker's stylish, colorful vestments. A 1997 article in *Forbes* magazine suggests that Hilfiger's 47 percent rise in earnings over the first nine months of its fiscal year 1996–1997 had much to do with the clothing line's popularity among the kinds of kids who shop Chicago's Madison Street.[4] Suddenly, Tommy Hilfiger became cool, not only among the urban teens but also among their counterparts in the suburbs. "That gives them a sense of pride, that they're bringing a style to a new height," Sarah Young suggests.[5] Thus, those who don't have much control over other aspects of their life find comfort in having at least some control over something—style.

There's another facet to this as well: the romanticization of urban poverty by some white teens. In St. Joseph, Michigan, a nearly all-white town of 9,000 in the state's southwestern corner, a group of teens mimicked the mannerisms and fashions of their neighbors across the river in Benton Harbor, Michigan, a nearly all-black town that has been economically devastated by the closing of the local factories and foundries. This cadre of kids called themselves "wiggers." A few white boys identified themselves with one of the Benton Harbor gangs, and one small band was caught carrying out holdups with a BB gun. A local police detective laughingly called them "wannabes." At St. Joseph High School, the wiggers greeted one another in the hallways with a high five or a twitch of the head. "Hey, Nigger, wha's up?" they'd inquire. "Man, just chillin'."

But it was through fashions—as consumers—that they most clearly identified themselves with their peers across the way. They dressed in the hip-hop fashion made popular by M.C. Hammer and other rap artists, wearing blue jeans big enough for two, the crotch down at their knees. (The beltless, pants-falling-off-hips style originated, many believe, in prison, where inmates must forgo belts.) The guys wore Starter jackets and hats, the style at the time. The girls hung braided gold necklaces around their necks and styled their hair in finger waves or braids. For these teens, the life of ghetto kids is edgy, gutsy, risky—all that adolescents crave. But do they know how edgy, how gutsy, how risky? They have never had to comfort a dying friend, bleeding from the head because he was on the wrong turf. They have never sat in a classroom where the desks are arranged so that no student will be hit by falling plaster. They have never had to say "Yes, Sir" and "No, Sir," as a police officer, dripping with sarcasm, asks, "Nigger, where'd you get the money for such a nice car?" From a safe distance—as consumers—they can believe they are hip, hip being defined as what they see in their urban counterparts. With their jeans sagging off their boxer shorts, with their baseball caps worn to the side, with their high-tops unlaced, they find some connection, though in the end it is a false bond.

It is as consumers that poor black children claim membership to the larger community. It is as purchasers of the talismans of success that they can believe they've transcended their otherwise miserable situation. In the late 1980s, as the drug trade began to flourish in neighborhoods such as Chicago's West Side, the vehicle of choice for these big-time entrepreneurs was the Chevrolet Blazer, an icon of suburban stability. As their communities were unraveling, in part because of their trade, they sought a connection to an otherwise stable life. And they sought it in the only way they knew how, the only way available to them: as consumers. Inner-city teens are eager to participate in society; they want to belong.

And for the white teens like those in St. Joseph, who, like all adolescents, want to feel that they're on the edge, what better way than to build some connection—however manufactured—to their contemporaries across the river who must negotiate that vertical drop every day? By purchasing, in complete safety, all the accoutrements associated with skirting that fall, they can believe that they've been there, that they've experienced the horrors and pains of growing up black and poor. Nothing, of course, could be further from the truth. They know nothing of the struggles their neighbors endure.

On the other hand, fashions in the end are just that—fashions. Sometimes kids yearn for baggy jeans or a Tommy Hilfiger shirt not because of what it represents but because it is the style of their peers. Those "wiggers," for example, may equate the sagging pants with their neighbors across the river, but kids a few years younger are mimicking them as much as their black counterparts. Fashions grow long limbs that, in the end, are only distantly connected to their roots.

Take that excursion down Madison Street—and the fault lines will become abundantly clear. One can't help but marvel at the spiritual distance between those shopping at Tops and Bottoms on the blighted West Side and those browsing the pricey department stores in the robust downtown. And yet many of the children have one eye trained down Madison Street, those on each side watching their counterparts and thinking they know the others' lives. Their style of dress mimics that of the others. But they're being cheated. They don't know. They have no idea. Those checking out the array of Hush Puppies at Tops and Bottoms think they have the key to making it, to becoming full members of this prosperous nation. And those trying on the jeans wide enough for two think they know what it means to be hip, to live on the edge. And so, in lieu of building real connections—by providing opportunities or rebuilding communities—we have found some common ground as purchasers of each other's trademarks. At best, that link is tenuous; at worst, it's false. It lets us believe that we are connected when the distance, in fact, is much farther than anyone cares to admit.

Oh, Isaac, Oh, Bernard, Oh, Mohan

Bharati Mukherjee

IN 1978, in Toronto, the owner of a kebab stall in a just-opened South Asian food court confided to me that his life's ambition was to set up roti-kebab franchises all over Canada and the United States. We were elbowing our way through a mob of fast-food eaters in party dress. It was a Saturday night, and whole Toronto neighborhoods and suburbs favored by Indo-Canadians (most of whom had immigrated during or after 1972) seemed to have emptied into that food court. The patrons ranged in age from those who needed walkers to those who needed strollers to get around. Families swirled around stalls, greeting friends and gorging on

cheaply priced combination platters. In the turmeric yellow light
and the fragrant air of that food court, satiety seemed synonymous with virtue. Like all virtuous patrons around me, I found
myself sampling disposable platefuls of pakora, samosa, kham-
man, dhokla, dosa, pilau, puri, naan, paneer matar, alu gobi,
chicken jalfrezi, mutton curry, and tandooried everything, including pheasants and prawns.

The kebab stall owner, an animated Indian immigrant in his
early thirties, also owned the food court building. He had made
his money through real estate investments. In a depressed white
neighborhood that included duplexes, stores, bars, a church, and
a Canadian Legion hall, he had bought a shabby arcade where
listless youths hung out and converted it into an emporium of
ethnic fast food. What might have started for him as just another
investment risk had, by opening time, become a social vision ful-
filled. That evening in the late 1970s when I met the owner, he
struck me as less a real estate speculator and more a proselytizer
for multiculturalism. The food court of his dreams was to be more
than a place to grab a spicy bite; it was to be a gathering place
where South Asian Canadians of all classes, ethnicities, and religions—Indians, Pakistanis, Bangladeshis, Sri Lankans, and members of Caribbean and East African collateral communities—
could gather as one. An island of Indian subcontinental multiculturalism in an ocean of Canadian multicontinental multiculturalism: in a Canada that boasted that its national policy of a cultural mosaic was more humane than the United States's melting
pot, the proprietor's goal seemed almost patriotic.

Before I left the food court that night, he declared: "I take my
cue from McDonald's. Roti wrapped around a kebab dog—why
not? In fact, how about kebabs stuffed into many different kinds
of Indian bread? I shall turn North America on to naan-kebab,
paratha-kebab, chapati-kebab! My roti going head to head with
tacos and pizza—why not?"

I visualized the neon arches of his roti-kebab franchises light-
ing up the whole continent. We're not talking art here, no third-
generation dream of writing the Great Canadian Novel; no gentle, self-doubting, roots-searching, familiarly guilty immigrants'

grandchildren's lament: "Why, oh why, didn't my parents force me to learn Hindi? What am I? Who am I? Who cares?" Maybe someday one of them, remembering her childhood running through the immigrant streets of Toronto or any of a hundred other Torontos, will turn out her own Indo-ironic version of Henry Roth's *Call It Sleep* or his own fondly bitter *Odessa Tales* à la Isaac Babel. Work for deep into the next millennium. This was the hard-core immigrant experience. Work the hours, invest the profits, shovel the snow, keep the shop open, gouge and hustle, ignore the scrapes and insults. (This is where my own immigrant writing experience was born, with the endless food courts, the sari shops, and the Indian grocery kiosks appearing as avatars from Bernard Malamud's *The Assistant*.)

The proprietor's optimism seemed almost modest as my husband and I walked the many blocks to where we'd had to park because so many carloads of patrons had converged on the food court. Those who couldn't find table space indoors had clustered around their vehicles, setting up take-out feasts on the hoods and roofs. Silken saris and brocaded kameezes brightened the gritty sidewalks. Satiated gourmands lingered on side streets to catch up on community news. The down-and-out neighborhood's only parking lot transformed itself into a raucous picnic ground. Hindi film music spilled out of lowered car windows. Families ambled from automobile to automobile, visiting or networking. Young mothers changed diapers on backseats. Men in dark suits exchanged tips on the stock market. Leather-jacketed and blue-jeaned teenage boys and girls slouched in segregated packs, showing off their cool.

The scene was uniquely New World, the mood euphoric: We've made it! Our hard-currency dollar buys more than unconvertible rupees! One more community of economic refugees was measuring success and happiness in terms of dollar power. That feeding frenzy inside and outside the food court was the community's self-celebration. A brown community was making it in white Canada.

The patrons were guileless, guiltless worshipers of materialism. They suffered no delusions about what had propelled them from an overpeopled subcontinent to a relatively barren one: better job opportunities; fatter incomes; a car or two in the garage of a sub-urban home on a rajah's-sized lot, which you bought on bank loan instead of paying for in full with life savings and dowry money at time of purchase (the reason why your father never realized his dream of owning his own home); good public education, available free for the children; and low-cost or no-cost high-quality health care for you and all your dependents. Those were just some of the basics that native-born North Americans considered their entitle-ment. But if you worked hard, saved diligently, and invested right, there was more to be enjoyed, much more—a Mercedes-Benz au-tomobile in your driveway; a big-screen television set, videocas-sette recorder, video camera, and sound system in your living room; a cell phone and electronic organizer in your briefcase. If you treated the New World right, it would make room for you at its table of conspicuous consumption.

The boom in the food court's sales did not bring a matching boom to the existing businesses in the neighborhood. Seedy hard-ware stores and failing beauty salons remained just that. Prosperity was community based, and this fed hostilities. The long-term residents saw themselves as victims of an invasion by New Canadians who were ignorant of, or unwilling to accept, Canadian codes of public conduct. Over the months, the more successful the food court enterprise became, the more vocal grew the complaints: too many cars, too many poor drivers and double-parkers; too much wailing cassette music in languages that were neither English nor French; too many curry eaters dumping foul-smelling garbage on the sidewalks.

Strategy sessions were held in the legionnaires' hall; petitions were delivered to the food court's proprietor, who had boldly dreamed of being North America's roti king; some compromises were reached. The proprietor bought more garbage bins and hired a security staff, at his own expense. The security staff did what it could to keep illegal parkers moving. But the extravagant ap-petites of patrons kept the bins overfilled and the sidewalks lit-tered with greasy take-out containers. Bollywood (the Bombay movie scene) sound tracks cranked up the decibel of nostalgia.

The complaints continued, some of them culturally sensitive and thus more difficult to finesse: Indians act like peasants; they leave gobs of spit and phlegm on sidewalks; they toss dirty diapers out of car windows. (What greater symbol of New World freedom than disposable diapers? I can sympathize with the gesture, that small release from baby-bondage that is one of Indian woman-hood's defining encumbrances. Still, not on the streets, please.)

Patrons had their own complaints against the neighborhood residents. Several South Asian families, independently of religion, level of affluence, or status within Canadian society, were as-saulted with spitballs and called "Paki! Raghead!" as they walked from their cars to the food court. There were near confrontations with white adolescents wielding hockey sticks. Young Sikh immi-grants, no slouches with field-hockey sticks themselves, orga-nized a flying squad of South Asian community defenders.

Mediators were called in, as race-related incidents were not uncommon in Toronto in 1978. Four years earlier, Prime Minister Pierre Trudeau's Liberal Party had floated a policy paper inviting a nationwide debate on what kind of racial mix and cultural jostling Canadians could tolerate in the "mosaic" of the future. It seemed to many of us that the desired response was preordained in the very phrasing of the questions. The catalyst for the debates had been the Liberal government's admission into the country of 2,000 Ugandan Asians holding British passports, all of them vic-tims of Idi Amin's expulsion order and all of them precipitously deprived of British citizenship through the introduction of the racist "grandfather clause." That humanitarian gesture had caused such a backlash among European Canadian voters that many of us South Asian Canadians personally experienced racial harassment in shopping malls, subway stations, and parking lots of fast-food outlets.

I sat in on just one of several mediation meetings, which was attended by neighborhood representatives, food stall owners, mu-nicipal leaders, and a member of the Ontario Human Rights Council. That meeting, as apparently had earlier ones, came to the conclusion that both sides needed to chill out before resum-ing negotiations. For me, that meeting was a visceral lesson in how—in these times of postcolonial diaspora—not just Toronto but every North American city was becoming freshly a frontier. As in any frontier town, the meeting hall was packed with winners and losers, stalkers of dreams and fugitives from nightmares. Listening to their impassioned speeches, I recognized on both sides idealists, hustlers, con artists, hotheads, fence-sitters, and drifters. But the lesson that took me by surprise was that a frontier town, with its insistent us-versus-them mode of self-defense, cannot tolerate the eclectic loyalties of an independent thinker. In that hall, it was my ethnicity, not my opinion, that designated me ally or adversary.

Many years later, after I had moved to the United States, become an American citizen, and settled in northern California, I came across Wallace Stegner's memoirs of growing up in a frontierscape of another kind. In *Where the Bluebird Sings to the Lemonade Springs,* he noted that

> there is something about exposure to that big country that not only tells an individual how small he is, but steadily tells him *who* he is. Any time when I lay awake at night and heard the wind in the screens . . . or slept out under the wagon and felt it searching among the spokes of the wheels, I knew well enough who, or *what,* I was, even if I didn't matter. *As surely as any pullet in the yard, I was a target, and I had better respect what had me in its sights* [italics mine].[1]

Stegner's "big country," though built up into cities with monumental skylines—cities often financed by immigrants and foreigners—remains a country of big potential. In February 1998, an India-born Californian, Sabeer Bhatia, the multimillionaire founder and chief executive officer of Hotmail, a company providing e-mail services, remarked to a reporter for *India Currents,* an ethnic Indian magazine based in San Jose: "This valley is a land of opportunity and this is the place where legends are created every day. I wanted to start a company of my own and pursue my dreams." He had just closed a dazzlingly lucrative deal with Microsoft and was being showcased, in the journalist's words, as "one of the youngest entrepreneurs to strike it rich and strike it big."[2]

Unfortunately, even in our overdeveloped and Web-connected "big country," Stegner's boyhood view of it as a place for armed predation and self-protection still feels true.

The last I heard from the owner of the Toronto food court, he was getting ready to start over in Florida.

The acquisitive thrill among South Asian immigrants is most sympathetically understood in the context of immigration history. Our dream chasing in the United States goes back only to the turn of the century, and our entry in any significant numbers in the all-American race for competitive consuming dates back only to 1965, when U.S. Attorney General Robert F. Kennedy replaced the traditionally racist region-based immigration laws with the granting of visas on the color-blind criterion of merit.

The first batch of Asian Indians known to have disembarked on North American shores were soldiers of the Sikh faith from farming villages in the Punjab. These soldiers had participated in Queen Victoria's Diamond Jubilee celebrations in England in 1897 and then, on their way back to India, had jumped ship off the coast of British Columbia. Over the next two decades, Sikh Punjabi villagers and a lesser number of Muslim laborers began to show up as farmworkers and loggers in the states of Washington and California. Although there were a handful of emigrants from the elite classes of urban India, among them civil rights activists Taraknath Das and Ajkoy Kumar Mazumdar, by 1920 most of the 6,400 South Asian entrants were uneducated, non–English-speaking migrants, economic refugees willing to work hard on railroads as well as on farms and in lumberyards for lower wages than American, Japanese, and Chinese workers.[3]

These brown Caucasian pioneers faced the institutional hostility of U.S. laws. In 1920, the California Alien Land Law and a series of United States Supreme Court rulings deprived them of the right to own land; in 1923, their eligibility for naturalization was retracted in *United States v. Bhagat Singh Thind*; and a year after that, future emigrants from the Indian subcontinent were denied

admission by means of the Immigration Act of 1924. In Stegner-
era "big country," where any "other" was seen as a threat to survival, the turban-wearing Sikhs were physically attacked by white
workers fearful of losing their jobs. There were what I can only
term pogroms against the Sikhs, the most brutal of which occurred in 1907 in Bellingham, Washington. The Asiatic Exclusion
League of San Francisco and the national media played on white
fears of racial mongrelization and dropping incomes, scapegoating
South Asian immigrants, whom, out of cultural ignorance or
media savvy, they labeled "the Hindoos"—in other words, an
unassimilable alien group of widow burners and sneaky mystics.

The physical intimidation and the legislated restrictions
worked. By 1940, there were fewer than 2,500 South Asians in
the United States, if census figures can be trusted. Many of them
were illiterate; almost all lacked a high school education; and because of antimiscegenation state laws and zero legal immigration,
a significant number of these "Hindoos" were actually of part
Sikh Indian and part Mexican descent.

A few closet Horatio Algers among the Sikh immigrants
learned to beat the discriminatory system by setting up white
sympathizers or entrepreneurs as the proprietors on paper of their
sizable orchards and farms. The majority of these economic
refugees, however, ended up poor, landless, familyless drifters
through the American dream–scape.

Robert Kennedy changed the culture, structure, and size of the
South Asian immigrant population when he pushed through immigration reforms: ask not only what America can do for the
trained immigrant but also what the skilled and professionally
trained immigrant can do for America. For the first time, merit
counted more than skin color. Back on the Indian subcontinent,
physicians, engineers, accountants, systems analysts, and computer experts—young family men who found themselves with too
much ambition and too few prospects in their overpopulated
hometowns—lined up outside the U.S. consulates for immigrant
visas. They were confident that if given a chance to "make it,"
they would.

And they did. Part of the success of this second wave of Indian immigrants had to do with their superior education and special-ized skills. The rest had to do with lucky timing. They were start-ing over in America just as the civil rights movement, the women's rights movement, and the anti–Vietnam War movement were managing to muzzle the more flamboyant public expressions of race-related bias. For instance, even though the majority of these newcomers were, in fact, practicing Hindus, by the mid-1960s anti-immigration agitators had dropped the use of *Hindoo* as a choice slur. Besides, unlike the Punjabi farmers of the 1920s and 1930s, who were technically British subjects and who there-fore Americans had tended to dismiss as an impotent, conquered people, these young families were emigrants from an indepen-dent, strategically located country. America's memory of its be-hind-the-scenes bolstering of British colonial rule in India in the early decades of the century (especially its crushing of the pro-in-dependence activities of the California-based Indians who formed the Ghadr Party) had been elbowed out by the flower children's fantasies about Eastern spirituality. In the 1960s, Americans tended to credit all Indians with having automatic ac-cess to wisdom and serenity.

This second wave of Indian immigrants accepted on faith America's promise that the talented and persevering go-getter is entitled to limitless material reward. But even more than being eager acquisitors, they were dedicated family men; what they planned to buy with their new dollar power was better schooling and thus better job opportunities for their children. The liberal-ized immigration laws of the mid-1960s had made it possible for them to bring their wives and minor children. The family has al-ways been the most defining and binding unit for South Asians. Back in the old country, multigenerational families (often includ-ing uncles, aunts, cousins, and village "brothers") traditionally lived under one roof. For example, I grew up in Calcutta in a household of forty-five blood relatives. In the new country, the family unit shrank from an extended one inclusive of all blood rel-atives to a nuclear one, making loyalty easier to focus and quality of family life less expensive to enhance.

Unlike the Sikh farmers of the first wave, a discriminated-against rural people who had tried to sustain the village lifestyle of Punjab, the newer, better-educated, city-bred arrivals were socially at ease in corporate and suburban America. Immigrant life was a bicultural interstate: you drove in the American lane at work and in the Indian lane at home. You adapted to American corporate manners in order to thrive professionally, but you resisted American culture, especially when it came to daughters dating. You pledged allegiance to the U.S. flag to guarantee that you would get your work permit. You didn't slash and burn the person you had been in your homeland; you just added a new improviser-survivor persona to the diversified portfolio that was your personality in order to manage unfamiliar risks. Now that you were oceans away from the hometown you'd escaped because it had stifled ambition, you could indulge in memories artificially sweetened with nostalgia. Only inept frontierspersons wore themselves out with cultural maladjustment.

Self-fulfillment assumed simple expression in the 1970s and 1980s: a house in the suburbs with at least one Mercedes-Benz in the two-car garage; investment properties that might include tenement buildings suitable for leasing to government agencies in need of single-room occupancy hotels to house low-income tenants; a diversified portfolio of stocks and bonds; healthy savings accounts in several different banks, especially foreign banks offering high interest rates; obedient, straight-A children bound for Ivy League colleges; and plans for a retirement life of easy luxury back in the homeland by means of real estate in one of many exclusive housing "colonies" reserved for paying-with-hard-currency nonresident Indians. What could possibly go wrong when your dreams were so uncomplicated and your skills so sought after in your (temporarily) adopted country?

I think now of the 1970s and the 1980s as the honeymoon period for my community of post-1965 South Asian immigrants. We had no personal memory of the anti-Asian prejudice, let alone of the pogrom-style brutality, that the earlier wave had suffered. Thanks to our luckier timing, we had entered a United States that had been macerated into mellowness by civil rights activists and Asia-philic nirvana seekers. Having grown up in a subcontinental culture that nurtured class and ethnic differences, we may even have resisted acknowledging how the "middle passage"—no matter how symbolic our variant of that passage had been—linked us, academics, psychiatrists, housewives, newspaper vendors, motel managers, Silicon Valley millionaires, Central Valley fruit pickers, to all immigrants in America.

In the 1990s, the South Asian community has expanded and its homogeneity of educational level and skill has been dissipated through the Immigration and Naturalization Service's favoring of family reunification. Over a Thanksgiving dinner for thirty poets and novelists from around the world in an Iowa farmhouse in 1997, a Bengali-speaking naturalized American with a lively sense of humor clued me in to some of the emerging class nuances. "Forget the Benz," she said with a laugh. "That's strictly for up-starts. Now the two must-have cars for the driveway are a Beemer and a Lexus."

A Boston-based gynecologist, she was visiting a younger sister who was finishing her doctoral studies in literature at the University of Iowa. Her husband was a radiologist at a hospital in Boston. The prospect of better opportunities for their daughter, an only child who was now a twenty-something professional, had tempted them to emigrate. The daughter had done well at a well-known boarding school in New England and then gone on to an Ivy League college. "It was expensive," the proud mother explained, "but we didn't skimp when it came to her education. The upstarts take shortcuts. They send their kids to private school just for the last two years of high school. That way, the parents save money without sacrificing admission to Harvard or Yale. For them, private school is just another smartly managed investment."

The Indo-American professionals are on to the way class works in this country that cherishes its national myth of classlessness. Almost unconsciously, they've had five thousand years of training.

Being smart consumers of the American class system, however, has had unanticipated consequences for many South Asian American parents. In some New Jersey towns where substantial numbers of affluent South Asians have settled, parents complain that their children are rejecting filial examples of ambition and diligence. In Rhode Island, an Indo-American student at Brown University described his teenage friends as "dropout wanna-bes." Whereas their parents have chosen to see themselves as largely irrelevant to the racial divide between blacks and whites and enlarged the Eurocentric American dream of the past century to include themselves, these U.S.-born children insist on claiming a space within that racial divide, looking to self-empowered minority communities, especially the African American one, for role models. That blinkered immigrant energy lasts just a generation or, at most, two. It's the grandchildren who will write the books.

If the parents are consumers of America's material resources, the children are nibblers of its ideological promises. The immigrants came to America in search of a better future, and now that future—their U.S.-born children, survivors of the high school culture of sensory experimentation and outspoken dissent—is consuming their traditional South Asian sense of self-respect. Children talk back to parents; some date on the sly; a few refuse the marriage partners selected for them, a still widespread practice, by their elders. A very few brides report their wife-battering husbands. It is more than a simple generational conflict. What I am witnessing, as an immigrant myself and now as a university professor in California with classes full of second- and third-generation Asian Americans who are beginning to write their stories, is the first and second generations of an Asian remake of the drama of Americanization.

I realize how deeply I participate in this three-generation drama. In the summer of 1997, in Prague, I served on a panel with an American writer of eastern European origin whose work I admire very much. We were presented as "ethnic American" writers. My copanelist made a familiar reference to the mindless, heartless, grubby mall culture of America—an understandable response to the preserved and restored splendors of Prague. I appreciated his observation, but I couldn't share it fully. My own immigrant experience with North American malls has been different. In states such as New Jersey, malls have become the places of choice to meet for immigrant housewives with a sense of self. Driving to the mall is itself an empowering adventure. Not

to be at home, cooking for husband and children? Not to be min-
istering to husband's mother? To be so selfish? Back in India,
most women didn't expect to earn a driver's license (and I never
have). To make appointments without consulting live-in in-laws,
to have friends who will waste a morning with you in public, is to
be free. And to keep those appointments in a consumer's heaven
like the Paramus Park Mall in Paramus, New Jersey, comes pretty
close to bliss.

Consuming Nature

Bill McKibben

To be under siege from a cloud of blackflies is to feel your sanity threatened. In and out of your ears they crawl, biting as they go; in and out of your nose, your mouth, the corners of your eyes. If you've covered up everything but your hands, they will start there and crawl to your wrists, leaving welts wherever they feed. I went out to the garden one spring evening without my shirt tucked in tight enough, and when I came in five minutes later my wife described to me the perfect row of bites, twenty or thirty of them, that ran along the narrow gap of skin that had winked open when I stooped to weed.

Blackflies hover in a cloud about your face and move with you
for miles, so great is their need for your warmth and company and
blood. Every writer of the mountainous North has tried to describe their voraciousness—"winged assassins," "lynch mobs,"
"jaws on wings." Here in the Adirondack Mountains of upstate
New York they constitute their own season, one that lasts as long
as spring or high summer or fall color (though not as long as winter). For six or seven weeks, from before Memorial Day to after
the Glorious Fourth, the paradise of a town where I live, an enormous expanse of mountain and river and stream and lake and
pond, is a paradise flawed. Most of the land here is protected by
the state constitution, proclaimed "forever wild," but the legisla-
ture has never managed to resolve away the blackflies.

It's not that no one's tried. As early as 1948, local towns seeking to extend the tourist season were spraying DDT from heli-
copters and tossing chunks of it into the streams. Rachel Carson
put an end to that by 1965 (and by the early 1990s the first ea-
gles were finally returning to the Adirondacks to nest, their eggs'
shells again thick enough to allow them to hatch). In subsequent
years, some towns used malathion or methoxychlor, sprayed usually from the air but always in the face of opposition. Then, more
recently, some scientists began experimenting with a more natural method of control, a naturally occurring bacteria called
Bacillus thuringinsis, which had been used for many years for organic control of garden pests. The *israelensis* subspecies, from the
deserts of the Middle East, is highly specific for mosquitoes and
blackflies. And so there was soon a small Adirondack industry of
private contractors who would bid for the right to treat streams
each spring, killing off the blackfly larvae in ways that appealed
to both environmentalists and tourist-seeking local businesses.

But our town had never gone in for BTI, as the treatment is
known, in large part because it is a frugal place, with the lowest
property taxes in the region. No one ever brought the question
up, and so spring after spring we had blackfly season, hard on the
heels of mud season. Then, suddenly, that changed. A petition
circulated demanding that Johnsburg join the list of towns that
treat their streams. The movement may have started one morning

at a Rotary Club meeting in Smith's Restaurant, at which a local realtor got up to complain that she'd lost a sale when she could not even get a couple from car to house, the flies were so thick. Sandy Taylor heard her and agreed to help write a petition.

Sandy Taylor and her husband, Jim, moved here not long ago from the South and before that the Midwest, where Jim had worked for the Monsanto Company for many years. They are exactly the sort of people who revitalize communities by moving into them. Before long Sandy was helping to organize our town's new library, the first in its history. The Taylors became mainstays of Rotary, of the church, of the theater group. They represent everything that is good about a certain American civic ideal, a spirit that is in many ways foreign to this backwoods spot. And it's not as if they are environmentally unaware or unconcerned; Sandy worked for many years as a guide at the biological research station run by Washington University in her hometown, St. Louis. "Our happiest memories as a family," she told me once, "are the camping trips we used to take."

But for her, as for most people, blackflies were not a desirable part of nature. "I can't garden, and I can't walk in the woods without all this protective paraphernalia, which is uncomfortable and hot and irritating," she told me. "My legs become a mass of bites that don't go away till August." Soon several hundred people had signed the petition she helped draw up, and the town board was busy drafting a set of specifications so it could put the job out for bids. Local innkeepers predicted that the cost might well be covered by the taxes paid by vacationers who would otherwise stay away. It looked like a done deal, as if our town would soon join the twenty-one other Adirondack communities that treat their streams with BTI.

Against most expectations, however, opposition began to form. It was not particularly organized—there was no official group, no "Save Our Flies" contingent. Instead, questioning letters started appearing in the local newspaper. Some of the comments concerned cost. "This is going to cost us $40,000, my share will be $56, and I don't even know if it's going to work," said one resident. Others questioned the effectiveness of the plans: Johnsburg covers a vast area, most of it deep wilderness, and since blackflies will migrate a good distance in search of the blood they need to lay eggs, all those streams would have to be treated, which some experts said was a dubious proposition.

But most of the opposition was unexpectedly philosophical. For one thing, the messages of thirty years of ecological thinking had begun to penetrate people's minds. The fact that there are millions of blackflies around Johnsburg in the spring, several residents pointed out, means that *something* must eat them for dinner. Fishermen testified that they had slit open trout bellies to find them crammed with blackflies; others worried about birds, or about bats, or simply about whether it was prudent to muck around with Such Vast Systems.

And there were the people who said, This is not such a big problem. Sure, a few days a year, when there's no wind, it gets bad, and so I wear my bug veil or I stay indoors.

And there was something more yet. A surprising number of my neighbors said—not always loudly, often a little backhandedly, maybe with a shade of embarrassment—that somehow the blackflies were a part of life here, one of the things that make us whatever it is that we are. Could we still have the Black Fly Ball at the local tavern, someone wanted to know.

I ONCE DID an odd experiment in which I found the largest cable television system on earth, which was at the time a hundred-channel operation in Fairfax, Virginia, and got people to tape for me everything that came across all the channels during the same twenty-four-hour period. I took my 2,400 hours of videotape home to the Adirondacks with me and spent a year watching it, trying to figure out what the world would look like were that one's main window on it. And what I found, amid the many lessons that spewed forth from the six home shopping channels, the four music video channels, the three sports channels, was this one overriding message: *You are the most important thing on earth.* You, sitting there on the couch, clutching the remote, are the center of creation, the heaviest object in the known universe; all things orbit your desires. This Bud's for You.

This is, of course, the catechism of the consumer society—the elevation of each one of us above all else. Sometimes it is described as "human nature," usually by people who would argue that you can't do anything at all about it. But of course in other times and other places, people have managed to put other things at the center of their lives—their tribe or community, their God, nature, or some amalgamation of these. Sometimes that's been all to the good: visit an Amish community. Sometimes it's meant pogroms. All I'm saying is that there have been other choices on offer.

Whether that still is true, however, I'm not sure. We have grown up in a culture so devoted to consumption—grown up so solid in the understanding that we define ourselves through certain patterns of consuming—that I doubt very much we can truly shake our conditioning. How else would we behave? From "real needs"? Save for the relative few of us who ever experience actual hunger or actual involuntary exposure to the elements, that sense of reality is as hard to summon as a sense of what it felt like to be chased by saber-toothed tigers. Poor people are just as interested in brand names as anyone else, just as devoted to the various cults (convenience, comfort, identity) of this central religion as anyone else.

And so it is no real stretch to say that the drive to eliminate blackflies from the small rural town where I live is simply one more manifestation of our deep consumer urge. We want to consume bite-free air; we want to consume our cedar decks and our pools and our gardens free of any complication or annoyance. We want to consume them *when* we want (not just on windy days) and *how* we want (bare-chested, with no damn bug veil). Jim Taylor spent the latter part of his career at Monsanto managing the AstroTurf division—managing the metaphor, fair or not, for conversion of the natural into the convenient.

BUT WHAT ABOUT those of us who oppose the blackfly treatment, we exemplars of biological virtue, eager to sacrifice ourselves for the sake of that great order Diptera and its thirst for our blood? How do we explain our escape from the great consumer faith into which we were baptized?

Mainly, I think, we do so by saying that we are just consumers, too. Why do I not want blackfly larvae killed in Mill Creek where it runs past my house? Partly because I don't want the biology of the stream tampered with but at least as much because I live not in Generic Suburban America, where everything is supposed to be convenient, but in the Rugged Frontier Adirondacks, where everything is supposed to be a challenge. At some level, I fear that I like blackfly season for the same reason I like winter and bad roads: because it heightens the adventure of living here. I consume inconvenience, turning it into a pleasurable commodity; it becomes the fuel for my own sense of superiority. I don't feel special because I own a particular brand of clothing, drive a particular make of car, smoke a particular brand of cigarette; I feel special because I have a crappy car, because I wear old clothes all the time, because it's a twenty-mile round-trip to get a quart of milk. I *like* it when people call up from the city to talk and the power has just failed, or a blizzard has just struck, or the temperature has gone to thirty below. I feel larger because of all that, I think; it pumps me up the way a Nike shoe, a Rolex watch, an in-ground pool, a Ford Explorer is supposed to pump us up. Blackfly season is a test, something to endure; I come out of it feeling tougher, stronger—which means, I think, that I'm a superconsumer, too. Blackfly season is about *me*.

And in this, I imagine, I am not alone. The shift toward voluntary simplicity now under way in some small corners of American culture is in some ways simply a shift toward a new self-image. Instead of defining ourselves by what we buy, we define ourselves by what we throw away.

There is clearly a sense in which this slightly submerged consumerism is more twisted than its straightforward counterpart. Elimination is a logical human response to blackflies, BTI a giant and efficient version of the timeless slapping hand. Wanting to consume fly-free air is, at some level, extremely logical. Finding a way to consume fly-*filled* air is more than a little nuts.

So IS IT ALL just a toss-up? If our is an age of endless irony, when nonconsumption is just another form of image building, does it make any difference how we live? Can you say that one path is better than the other? Can you say we shouldn't kill all the damn blackflies?

You can, I think, though you have to say it carefully, aware that your own sense of superiority is more than a little absurd.

The first argument is clear: even if the main reasons why you defend blackflies or recycle your dental floss have to do with you, they nonetheless benefit the rest of creation. Whereas normal consumption is almost by definition costly to the earth, this more rarefied form is almost by definition cheap and undamaging. This is a great practical virtue, since the results of normal, everyday consumer life now threaten to wreck everything around us. I've spent much of the past ten years writing about global warming, which is nothing more than the sum total of our lavish devotion to convenience, comfort, and power transmuted into several extra watts of solar energy per square meter of the earth's surface. It is human desire translated into planetary physics, and unless we can get those desires under some kind of control, the physics will turn impossible. By this analysis, though it may be bizarre to consume by not consuming, doing so is like supplanting heroin with methadone; one's cravings are stilled with minimum damage to the underlying system.

And yet there is something more to it than that. By its very nature, this kind of somewhat silly nonconsuming puts us in harm's way—raises the possibility that we will be exposed to forces that might actually change us, might begin to erode some of the conditioning we've carried since near birth. An example: When I lived in New York City, I helped start a small homeless shelter at my church and spent many nights there. This was classic non-consumer behavior, robbing me of many hours I might have spent in restaurants, bars, movie theaters, and boudoirs. But of course I did not do it primarily because I was a good Christian; I did it because I wanted the sense of being a slightly sainted fellow. Over time, however, the mere fact of being there began to change me in certain small ways. I learned that in some fashion it made me feel peaceful to do the small daily tasks of that place—changing the sheets, cooking the soup, delousing the pillowcases. It was one of the paths to learning not to resent housework, one way to cease the innate consumer desire for a maid (or a mother). In fact, I sensed, counterintuitively, that this work made me happy—a revelation that would not have surprised any of the long chain of gurus and Christs and other cranks down through the ages but certainly shocked my suburban soul. Having been exposed to some deeper (if transient) joy, I was marginally less of a sucker for the various ersatz appeals of popular culture.

Sometimes now I help with the campaign to return wolves to the Adirondacks. They were wiped out here in the last years of the nineteenth century by people who thought of them in the way realtors now think of blackflies—as an annoyance standing in the way of progress. I try not to pretend to myself, any more than I have to, that my main interest is with the wolves themselves or even with the health of the forest, which badly needs a top predator. I know that what I want is to hear a wolf howling in the woods because it will make this place, and my life here, feel yet more romantic. I will consume that wolf howl, just as my predecessors consumed the quiet of their suddenly wolfless nights. But once the wolf is there, its howl will also carry certain other, less obvious messages; and there will be the remote chance of an encounter with this other grand representative of creation, an encounter that might go beyond mere consumption. I saw a grizzly bear one recent summer in Alaska, not far away on a muddy bank on a foggy night, and the sheer reality of that encounter shook some small part of me out of the consumer enchantment into which I was born.

Blackflies accomplish this, too, in a subtler way. They remind me day after day in their season that I'm really *not* the center of the world, that I'm partly food, implicated in the crawl and creep of things. They are a humbling force, and even if for a time I can involve them in my self-aggrandizing myths, they still exert a slow and persuasive pressure of their own. Over the course of a decade, living in a place dominated by high mountains, wild winters, summer storms, trackless forest, and hungry insects has in fact warped me in certain ways. I am not the same person who came here. I am still a consumer; the consumer world was the world I emerged into, whose air I breathed for a very long time, and its assumptions still dominate my psyche—but maybe a little less so each year. And perhaps they dominate my daughter just a little less than that. There are times when I can feel the spell breaking in my mind—the spell of the advertiser on the tube, even the spell of the mythmaker in my mind. There are times when I can almost feel myself simply being.

At least for this year, Johnsburg decided not to use BTI. Instead, a questionnaire is being sent out with the tax bills. If the town were to treat the streams, it asks, would you be willing to give the workers access to your land? I think quite a few people— by no means a majority, but probably enough to make the plan unfeasible—will say no. Like me, they'll probably do it without quite knowing why. But it's one small sign for me that the enchantment is wearing off, that the incantation sung over our cradles by the television set may be less permanent than some think. A sign that spring may be coming—and with it the biting flies, by God.

A News Consumer's
Bill of Rights

Suzanne   Braun   Levine

SOMETIME IN THE LATE 1980s, readers of newspapers, viewers of evening news broadcasts, and other citizens who wanted to keep up with current events were all transformed into consumers. It was all quite imperceptible at first—a few more MBAs on the "business side"; a few special sections for special markets; a few focus groups and marketing strategies—and before they knew it, every audience member had become a "target." As such, they were courted and polled for interests and biases and profiled into demographic niches. The information gathered by the trend-mongers enabled news operations to find ways to entice the audience to their product—formerly the news—and to entice advertisers to their audience.

Some ten years later, the big word in publishing and broadcast journalism is *branding*. What used to be called the editorial product has been refined down into bite-sized media units that can most efficiently be marketed from the printed page into cyberspace or from the airwaves into a theme park promotion, as well as to the consuming public (formerly known as citizens).

News generated by this system often seems unworthy of the grandeur of First Amendment protection, and the citizen-consumers often seem to be complicit in the tarting up of the process that the Founders were sure would guarantee the well-being of the Republic.

While citizens were being downgraded to consumers, journalists were being demoted, too, from hero to villain. The adversarial stance that was seen as courageous during the Watergate scandal degenerated, in the eyes of the public, into gutter-style bullying; reporters came to be seen as mad dogs out to tear down big shots, build up celebrities, and make fun of everyone else.

Indeed, the media have betrayed the public trust in many ways, and the accusations that come their way are intense, if not always well founded. Journalists are blamed for sadistically sticking microphones into the faces of suffering people who are caught in a news event and for the cynical tabloid slogan "If it bleeds, it leads." They are blamed for the circus atmosphere created by the presence of cameras in the courtroom, which many think undermined the O. J. Simpson murder trial and, more recently, the trial of British nanny Louise Woodward. They are blamed for making heroes of such pillars of the community as Amy Fisher and Joey Buttafuoco and driving away potential leaders such as General Colin Powell. A statement in deputy White House counsel Vincent Foster's suicide note, "Here, ruining people is considered sport," was understood by everyone as referring to the Washington press corps. A 1994 poll conducted by the Pew Research Center for the People and the Press nailed the point grimly: more than 70 percent of the respondents said that "news organizations get in the way of society solving its problems."

The public, goes the response from media executives, should not complain; they are only getting what they have made clear in the marketplace that they want. Is the moral of this tale that the sordid press and the vulgar public deserve each other?

Surely, there is a better self lurking in a press corps that is currently trusted no more than are used-car salesmen, and there must be a better self in a public that is defined by its lowest common denominator. As things stand, some good impulses are being mobilized, but most are going awry.

The most dramatic venue the public has found for punishing the media for their sins is the jury box. When the Food Lion supermarket chain was awarded $5.5 million for damage to its reputation caused by a story based on questionable reporting practices, the real news—the shocking health violations and consumer fraud the undercover reporter had revealed—was lost in the shuffle. The consensus was that the media had been put in their place.

When the public raised an indignant outcry against the press for stalking Richard Jewell, the man falsely accused of the bombing at Centennial Olympic Park during the 1996 Summer Olympic Games in Atlanta, they were protesting arrogant, intrusive, and deceptive journalistic behavior across the board. But in fact, the smoke of their burning outrage obscured the very important news that it was the Federal Bureau of Investigation, not the press, that had behaved improperly, by leaking Jewell's name as a suspect for reasons of its own.

In both cases, issues of real public concern—issues of genuine civic urgency—were obscured by misguided revenge against the press (and, yes, by sloppy journalism).

This diffuse hostility on the part of the citizenry and these reprehensible lapses in standards on the part of the media come at the worst possible time in terms of the public's stake in legitimate news reporting. Two outside forces are battering at the gates. The first is a deluge of undigested, unverified, unreliable information being beamed down from cyberspace; to make responsible sense of it requires the kind of safeguards and values that journalism has developed. The second powerful force is building with every media conglomerate that is formed: a growing potential for control of the news.

Citizens who are consumers of news have a lot at stake here, and journalists, whose mission is to engage and serve those citizens, have a lot at stake in ensuring that consumer demands be stringent but realistic. To that end, both sides need to come to an informed understanding of which parts of the problem are a function of press performance—and therefore are fixable by the press—and which are the result of a host of other societal and economic factors and therefore need to be fixed in other ways (including by a crusading press). The beleaguered media community would surely agree with the New York–area discount clothing chain that advertises, "An educated consumer is our best customer."

To begin this very tricky education process, it is necessary to review some journalistic principles that are rightfully venerable. They establish the ideal of the profession, but they are not enough: when applied to the real world of news gathering, they are too lofty to be helpful in the moment when a crucial judgment call must be made. A workable News Consumer's Bill of Rights must offer a code that reflects an ethic that is both pragmatic and honorable.

The generally accepted principles of responsible news gathering are, roughly, as follows:

> To be fearless and compassionate—the modern corollary of the muckrakers' motto, "Afflict the comfortable and comfort the afflicted."

> To gather all relevant facts and assemble them into the truth.

> To be fair—to give all sides a chance to be heard.

> To be objective and unbiased.

> To inform by collecting and organizing information in such a way that it is helpful in making good judgments.

> To be meaningful—to ensure that the choice of stories to pursue is wisely made and, by the same token, that what is omitted has been thoughtfully evaluated as well.

To be responsible—if a mistake is made, to correct it quickly and effectively.

There are also a few things that everyone agrees ethical journalism should *not* be and should *not* do:

Above all, it should not be motivated by self-interest, commercial interest, or any agenda other than informing the public.

It should not incite violence—as in shouting "Fire!" in a crowded theater or shouting "racism" in a tense confrontation.

It should not invade the privacy of any individual, even a celebrity.

To illustrate how news values work in the field, imagine, for a moment, a piece of news before it is filtered by the press: a friend calls to say that a woman in the community has been raped. The typical reaction would be, "Oh, my God!" followed by a flood of questions. "Who was raped?" "No, I don't know her; where does she live?" "Where does she work?" "Was she beat up? Who did it? Did they catch him?" And, if not, "What did he look like?" Any curious and possibly endangered citizen would feel entitled to the answers to these questions—the public's "right to know."

Now, insert a journalist and some news values into the picture and see what happens. The reporter will have had the same initial reaction of shock and concern, but she knows it is her job to put those feelings aside and tell the story in an objective, cool-headed way. She will round up as many facts as she can, facts that would answer all the questions the concerned citizen asked, but she will probably not be able to use them. If, for example, she works for a news organization that has a policy of protecting the privacy of rape victims, the reporter will not be able to reveal the victim's name, address, or place of work. Good taste might dictate that the reporter not reveal gory details of the assault or the nature of her injuries. (Unfortunately, the reverse is just as likely: the reporter's boss might be pushing for the most lurid details.) And as for information about the perpetrator, the police may not want his name revealed until they have him in custody. Moreover, even though a description is not very helpful if it doesn't include race, perhaps local black leaders have called attention to a pattern they have observed in which the only black people who show up in the news are criminals, and newsrooms in the community have become sensitive to the charge. So what is left after the story has been vetted for publication? Many fewer answers than the informative neighbor was able to provide.

Sometimes members of the media find themselves forced to make judgments that have more to do with their being in the middle (*media*ting) than with the news product itself. That is what happened in Waco, Texas, back in 1993. Members of a cult called the Branch Davidians and their leader, David Koresh, barricaded themselves in their compound in a face-off with a small army of federal agents and what appeared to be more members of the press than Waco had citizens. The actions of those journalists literally changed the story they were covering.

For starters, it was a well-intentioned member of the press who tipped Koresh off to the impending raid. A reporter whose newspaper had gotten wind of the secret FBI plans saw a mail carrier heading toward the compound and warned him that he was walking into danger. The reporter must have believed that he was making a moral decision to protect an innocent bystander; what he did not know was that the mail carrier was a member of the cult. Instead of turning back, he rushed ahead to warn those inside.

The fact that an employee of the local paper broke the news of the raid despite law enforcement requests to hold the story is still reverberating. In 1996, the families of the officers killed in the raid won a lawsuit in which they charged that by undermining the surprise element of the attack, the newspaper had contributed to the deaths.

It wasn't only print journalists who made decisions that turned out to have life-and-death outcomes. A local radio station made the decision to enter into the negotiating process by agreeing, as part of a proposed resolution of the standoff, to put Koresh on the air. At exactly the same time, CNN was working on the story and making decisions of its own, and when Koresh called CNN, he was put right on the air. So the negotiating team lost the bargaining chip of access to the national audience.

What is harder to pinpoint is the degree to which the media saturation of the area contributed to the escalation of tension. Would Koresh have given up if his worldwide platform had been removed? And by the same token, would the FBI have moved so precipitously if it had not been under pressure from the watching world to "do something"?

Which of the events at Waco might have been contained had journalists adhered to one or another item on the list of principles we started with? Were any of the events exacerbated by a decision based on one of those criteria? The point is that although there is much to aspire to in the lofty principles of truth-telling, independence, respect for privacy, and so on, these principles are little help to those on the ground directing coverage of a breaking story.

Take one last example. In 1992, tennis star Arthur Ashe was forced into a public announcement of the fact that he had AIDS by a telephone call from a reporter for *USA Today*. At the time, many people felt that he was "outed" for no other reason than to feed the public's insatiable appetite for gossip. Others argued that the revelation made people pay more attention to the disease because a national hero who was heterosexual had it. Ashe himself later said that it was somewhat of a relief to get his secret out in the open and that he believed it enabled him to do some good in the community in the months before he died.

Should the reporter have buried the story—even at the risk of seeing it reported elsewhere, with another reporter getting credit for the scoop? Maybe so, but what happens if the celebrity who is sicker than is publicly known is a presidential candidate—Paul Tsongas? Or a president—Franklin Roosevelt? Doesn't the public have a right to know?

But does the public have the right to know about a president's failing health even if the nation is at war and the news would give an advantage to the enemy? The most common justification for press censorship is national security. It would certainly have been treasonous to reveal plans for the Normandy Invasion of World War II. But what about the Bay of Pigs invasion of 1961? When President John F. Kennedy learned that several news organizations had gotten wind of the planned invasion of Cuba, he persuaded them to hold the story, in the national interest. But later, he lamented that if the story had gotten out, the ill-fated invasion would have been aborted and he would have been saved from a fiasco.

No two stories are alike, and each judgment call involves a slightly different mix of exposure, risk, and sensitivities. Sometimes the "watchdog of democracy" acts more like a pit bull, but the greater danger is to muzzle the beast. Nevertheless, a consumer of news is entitled to demand that editorial decisions are made in an honorable and informed environment, with safeguards and checkpoints in place up and down the chain of command to guard against carelessness, meanness, and misinformation.

That chain of command has more links than many news consumers are aware of, and it is important to keep in mind that in most cases, it is not the reporter but the editors and other supervisors who determine what is a "good story." Behind them are increasing layers of interests, which can extend up to the highest echelons of a corporation that may have only a passing interest and little expertise in news gathering.

For many such corporations, the only measure of success is the bottom line. Cutbacks and economies are management's goals. Journalists sometimes feel like dime-a-dozen field hands. Even in the better newsrooms, fewer reporters are covering more stories and accumulating less expertise than in the past. As they race from the latest floods along the Red River to a new AIDS drug, they take shortcuts, working, for example, from press releases issued by interested parties, telephone interviews with people in authority, and the reporting of others.

This slipshod approach not only undermines the quality of legitimate news reporting but also weakens the barrier to the kind of information the public most needs protection from—undigested rumors from the Internet grapevine. To understand how significant a player the World Wide Web has become, consider the fact that in the first month after the *Pathfinder* landed on Mars, the National Aeronautics and Space Administration's Web site had 566 million "hits." That statistic is a challenge to the best of new media, but the Web is just as often the source of misinformation, sometimes assembled into elaborate conspiracy theories. The "friendly fire" explanation for the crash of TWA Flight 800 crash is one: it was picked up by former presidential press secretary Pierre Salinger, and despite the efforts of investigative journalists and agencies that found no corroboration, Salinger, egged on by Web conspiracy buffs around the world, wouldn't let go.

Second-rate reporting is one by-product of corporate gigantism. Another—harder to recognize because it affects what *does not* get reported—is an environment of self-censorship, the common understanding among news staff members that some stories are off-limits. Talk to any journalist who works for one of the media monoliths and she or he will tell of having been discouraged from pursuing certain leads by editors and news directors who are trying to second-guess the titans they work for.

Rupert Murdoch is one. His empire includes newspapers and networks that cover politicians around the world who, in turn, make decisions that affect his business. He also owns book-publishing companies that can give hefty advances, which go way beyond projected sales figures, to powerful people who affect his business. And he owns movie companies that make films, such as *Independence Day,* that can promote his other media operations. In that movie, a news agency covering an alien invasion is called Sky TV. That happens to be the name of Murdoch's satellite operation, which had not long before pushed the venerable British Broadcasting Corporation out of the Asian market by—wheels within wheels—building "trust" with regional political power brokers, who, needless to say, expect to be treated respectfully on Sky News. How the Los Angeles Dodgers, which Murdoch bought in 1997, will fit into the family remains to be seen. But fit they undoubtedly will.

There is evidence that as the media have been absorbed into the big-stakes business world, money has become a more compelling interest than the public's right to know. The giants are particularly craven before the newest weapon in the litigation game: the multimillion-dollar lawsuit. Judgments such as that in favor of

Food Lion and against ABC (owned by the Walt Disney Corporation) undoubtedly had executives at news operations everywhere thinking twice before authorizing risky undercover reporting. The very suggestion of the possibility of a lawsuit from a cigarette manufacturer was enough for the CBS Corporation (which was in the process of acquiring a new corporate owner at the time) to cancel a controversial *60 Minutes* segment.

With such monumental powers at play, powers that are clearly beyond the control of the community of journalists on which the public has been focusing its wrath, is there any group or entity that news consumers can hold accountable? The first line of attack in any consumer action is the marketplace, and a newspaper that loses circulation or a television news show that loses Nielsen ratings will feel the vote of disapproval, but the response may not be a change for the better. To improve the news-gathering process, the public needs powerful allies on the inside. And vice versa. Many journalists are as unhappy with shoddy manufacturing practices and second-rate product as the public is but believe they have few supporters. Together, the public and the press could put the squeeze on the system to do better.

An example of such an alliance took place in Chicago in the summer of 1997, when citizens, community leaders, and members of the press joined forces to protest the compromise of news programming. The occasion was the appointment of tabloid talk-show host Jerry Springer as a commentator on the local NBC affiliate's news broadcast. Viewers let the station know that they deplored the legitimizing of a media personality whose claim to fame is his ability to create forums on varieties of degradation. But it didn't end there. Civic leaders who had been regular sources for comment announced that they would not be available as long as Springer was on the news team; and then, in a most dramatic gesture, Carol Marin, the popular and well-respected coanchor of the show, quit. That did it. Springer had to resign. Marin was ultimately rewarded for her principled stand and moved on to the news department of another Chicago station,

WBBM, a CBS affiliate, and Springer went back to delivering his brand of entertainment with continued success. News executives at WMAQ and around the country were put on notice that lapses in judgment and quality would not be tolerated.

To build this kind of alliance, citizens and journalists need a common standard to rally around, one that reflects a consensus not only on the goals of responsible journalism but also on the rules of combat. Although journalists are rightly blamed when they display an indiscriminate killer instinct, they should not be blamed for seeing their calling as a fight—a fight to get information that is not always forthcoming, to penetrate barricades of lies and resistance, to battle bureaucracies and bullies for the truth, and, in no small part, to battle their bosses to allow them to tell the kind of story they think needs to be told.

Fairness is a value that has been enlisted on so many sides of First Amendment controversies that its meaning has been diluted, but it can be focused to function as a guiding principle in the struggle to get the story to the public. Suppose the objective of fairness were understood to lie beyond reporting on the pros and cons of an issue or story and beyond the treatment of the people involved; suppose the primary mission of news reporting were to be *fair to the public*. Thus, a dedicated journalist would ask not only whether the story is being well and professionally told but also whether the public is being fairly served in the telling. Surely the public would buy that notion.

The fairness test would be applied this way:

It is clearly not fair to the public if, for example, the news is overrun with crime stories, to the exclusion of most others (for the average local television news program, the ratio is two to one) and if complex events are given only simple headline treatment ("More about late-term abortion after these messages").

It is not fair if the press squanders its credibility by cooperating in the blurring of fact and fiction or of news and entertainment.

It is not fair to turn journalism into propaganda, but by the same token, it is not fair to either the public or the press to adopt a standard of objectivity so rigid that it obscures the truth as the reporter sees it. Balanced reporting should not mean finding the same number of experts to defend Hitler as to condemn him. Good reporting is thorough and inclusive, and it is also informed. If journalists are the people's surrogate investigators, it only stands to reason that what the reporter has understood is as much a part of the story as what he or she has found out. At best, given the immediacy of journalism, any story is only what it looked like to that particular journalist at that particular time, measured against that journalist's accumulated expertise and within that journalist's ability to think, to find out, to tell.

Most unfair of all is to define news as bulletins about conflict between opposing forces: good guys and bad guys, winners and losers, horse races, power plays. What this approach does to important public issues—such as abortion, gun control, and affirmative action—is polarize them, moving compromise and resolution further out of reach.

There is a significant physiological consequence to the drumbeat of conflict, danger, and violence. According to psychologist and reporter Daniel Goleman, a human being confronted with news that is overwhelmingly threatening goes into primitive fight-or-flight mode: adrenaline takes over, and—this is the crucial part—the brain shuts down to survival level, eliminating any capacity for nuanced thinking. It is no surprise, and it is clearly not fair, that news consumers come away from most news experiences with adrenaline pumping and no sense of having learned anything.

The straight-to-the-gut format works only for the narrow universe of events that can be reduced to nicely matched opposites; the rest rarely make it onto the news radar screen. Thus, three important categories of information get short shrift or, worse, get mangled in the process of being squeezed into the combat mode: stories about process rather than outcome; complex stories that require the assimilation of material from various sources and the making of subtle connections; and stories whose import is an idea rather than a headline event. The annual budget debate, for example, could be covered from any or all of these perspectives, but instead it is trivialized by conventional coverage. Told in sporting terms as a tale of congressional huddles, fakes, and blocks, it is reduced to a series of battles, each one a tiresome replay of previous ones. Told in terms of individual items, it is reduced to a dull, dull, dull litany of incomprehensible billions of dollars. But as Gloria Steinem has pointed out, the federal budget is the nation's only statement of values. As such, it is a major story that would explore the real meaning of dollars earmarked for military, the arts, roads, children.

Likewise, when religion comes up, it is most often within the familiar parameters of scandal (priests who abuse children get a lot of coverage) and politics (the religious right, the pro- and anti-abortion crusades, the potential influence of the Promise Keepers). Questions of life and death—the basis of much of Western thought—are addressed, if at all, in the latest installments of the ongoing cat-and-mouse game between Dr. Jack Kevorkian and the law.

Sound bites cannot take the place of discourse when so many of the problems Americans confront are complex and messy and are not resolved easily, if at all. And gut-wrenching accounts of personal tragedy cannot take the place of a forum for the evaluation of collective solutions. (For example, of all the stories about individual child-care arrangements that were generated by the "Boston nanny" trial, not one addressed the fact that the United States is the only industrialized democratic country that has no national system of childcare.)

When thinking isn't part of what the news is selling, the public is not being fairly served. Situations that call for societal change may produce stories that begin with conflict, but in the middle they are about ideas, experiences, experimentation, and compromise, and the end is rarely a neat wrap-up. Recognizing this, New Jersey's governor, Christine Todd Whitman, proposed that the press join with politicians "to create an atmosphere that will allow people to float ideas." As things stand now, she lamented, "debates take the form of public slugfests; compromises are considered retreat." Something drastic—something as unimaginable as an alliance between the media and the politicians—needs to be done to "improve the tenor of public discussion."[1]

Alongside that improbable alliance, the notion of a mutual interest shared by news consumers and news gatherers seems totally reasonable. Working together under the banner of fairness to the public, they just might be able to co-opt the crass brand-consumer relationship and focus on reforming the very system that is profiting from it. The educated news consumer would be responsible for evaluating the product, with enough expertise to know where to look and whom to hold accountable; like any activist consumer, this one would be on the alert for shoddy manufacturing practices, misleading advertising, and dangerous products and would also take responsibility for spreading the word when first-rate goods and services are found. If journalists know that the public is doing its job, they can do their work, confident that there is a market for high-quality merchandise. This two-pronged attack on junk news is the best hope for reinvigorating the compact—sealed by the First Amendment—between news producers and news consumers.

When We Devoured Books

André Schiffrin

Books REMAIN an important part of American life. In spite of television and the movies, in spite of the Internet and the declining readership of newspapers; in spite of the dramatic decrease in the number of independent bookstores and the change in ownership of major publishing houses and book clubs, people still read. Some books still sell in very large numbers, though sales are not as large in relation to population as was the case with some of the books that captured the public imagination in the preceding century.

Indeed, if we look back to the nineteenth century, we see some revealing and sobering points of contrast. Although all too little research has been done on reading habits and book sales in the United States, we do know something of which books were the most popular in our public libraries 100 years ago. And what we know might seem surprising: in spite of the availability of a great many tawdry "popular" books, the most-requested titles were by such authors as Sir Walter Scott, Charles Dickens, and Honoré de Balzac.

Today, you need to search long and hard to find a best-selling literary translation. Most of the larger publishers have given up on that category altogether. Likewise, books containing new and controversial political ideas could sell in the nineteenth century what would be the equivalent (again in relation to population) of millions of copies today. Edward Bellamy's *Looking Backward* and Henry George's works advocating the single tax were among these. What's more, they were not only best-sellers. They also spawned discussion groups and political movements. Very little in recent decades has sold nearly as well as those books or the political best-sellers of the 1930s and 1940s.

The sales of these books fly in the face of the argument that American book publishing has lacked a substantial audience beyond the middle class. It's striking to note that when *Uncle Tom's Cabin* was published, California gold miners would pay two bits per night—quite a lot of money in those days—to borrow the book and read it after the day's labors. Do we even know these days whether workers read at all and, if so, what appeals to them? Harriet Beecher Stowe's book, along with the others just mentioned, was truly national in its appeal, transcending class and regional differences.

Why does this seem so improbable now? What has happened to the use, the consumption, of books since that time? I don't think Americans have changed so much that they no longer have the capacity to read and enjoy good literature or to be interested in serious books about what is happening—and what should be happening—in society. Rather, in the world of books, as in so many other cultural fields in America, choices have become increasingly limited. The "commercial" books are easier than ever to find—every airport newsstand and chain bookstore has the latest blockbusters. But the broader selection of serious literary and intellectual work that used to be widely available is now hard to find. Many of the publishers once known for such books no longer even bother to keep them on their backlists.

Perhaps the best way to understand the changes I'm describing is to look at the most popular book format in America, the inexpensive paperback. This format was pioneered in the late 1930s by Pocket Books, with its ever present kangaroo logo; by Bantam Books; and, later, by Penguin Books. A look at the early lists of those firms is instructive. One would expect them to consist mostly of westerns and mysteries, and indeed there was no lack of books by Erle Stanley Gardner. But alongside such best-sellers as James Hilton's *Lost Horizon* and Whittaker Chambers's translation of Felix Salten's *Bambi* one also found *Martin Eden,* Jack London's radical classic, now unavailable in any edition. At the same time, readers could buy Margaret Mead's *Coming of Age in Samoa,* Marquis Childs's *Sweden: The Middle Way,* and a long list of similar titles.

So much a part of American life were these books that the federal government distributed millions of them, free of charge, to soldiers during World War II. Few countries can boast that reading material went out along with the K rations, free to whoever wanted it. Indeed, the noted German critic and essayist Hans Magnus Enzensberger writes movingly of his discovery, when the Americans took over his part of Germany, that the soldiers had cases of these precious volumes. He had never before seen such books, and they became the basis of his own education in English and in non-German thought.

My first publishing job was with one of the country's biggest mass paperback houses, the publisher of the nonfiction titles I've just mentioned, called the New American Library of World Literature, Inc. It was the successor to Penguin USA, and its slogan was "Good reading for the millions." It was clear, at the time, that many people really were in publishing for that reason. I remember sitting in on editorial meetings and hearing the editors' serious discussions of how to present new and very demanding work to a mass audience. Of course, the New American Library published a vast array of westerns, mysteries, and the like. But it also published all the works of William Faulkner—not to mention Curzio Malaparte and the Italian postwar realist Pier Paolo Pasolini. Like the nonfiction titles mentioned earlier, these literary translations now can be found, if at all, only in very expensive small press reprints or university press editions.

These paperbacks cost twenty-five to thirty-five cents and could be bought at newsstands and drugstores throughout the country. With inflation accounted for, they would cost about $2.50 to $3.50 today. The price of a pack of cigarettes was more or less the rule of thumb for what a book should cost. One of the most expensive books we published was James Farrell's *Studs Lonigan* trilogy. It was so long that we had to charge fifty cents for it. The marketing people finally decided that the book's spine should be broken into two bands so that people could see that they were getting two books' worth and wouldn't feel cheated.

The covers of the paperbacks from that period were uniformly lurid; if you didn't look at the title, it would be hard to know whether what you had in your hand was by Mickey Spillane or by Faulkner. But there was a real attempt to get a broad audience to read the best that was being published. Even if Faulkner was described in each of his books as the author of *Sanctuary* (presumably the one "dirty" book that many people had the chance to get their hands on during adolescence), nearly all of his work was available. It would be many years before his work would become a staple of college courses, ironically losing most of its popular audience as it became elevated to the canon.

New World Writing, a very esoteric paperback literary review, was started at that time. To give a rather extreme example, one issue that springs to mind offered a selection of contemporary Korean poetry. Again, though, this was published for a very broad audience, with initial printings of 50,000 to 75,000 copies. *New World Writing,* whose name was derived from John Lehmann's famous wartime *New Writing,* published by Penguin (U.K.), carried on with that publication's popular-front attitude toward the masses and literature: the belief that ordinary folks could read the best in American culture and ought to be able to find it in every drugstore.

We didn't see the readers as segregated into an elite public and a mass public to which one had to pander. Even if you were publishing novels by Mickey Spillane or Kathleen Winsor's *Forever Amber,* you were also supposed to try and publish the best work you could find. And we expected that people would actually read it. Moreover, even if you thought something might be a huge seller, there was a limit to the pandering to popular taste, which kept you from publishing pornography and books that you believed were demeaning to the human spirit—books that now flood the market.

How did these values come to be replaced by the now dominant market values? My own experience is revealing of the forces at work.

After my stint at the New American Library, I joined Pantheon Books. It was 1962, and Random House, which had just gobbled up Alfred A. Knopf, had recently bought Pantheon as well. In those days, Random House looked like a giant in American publishing, but even with its two newly acquired firms, its total sales were only about $15 million per year. The Random House group now sells about $1 billion worth of books each year, and its owners, Donald E. Newhouse and Samuel I. "Si" Newhouse Jr., are said to have assets of $10 billion, with Random House only a small part of their overall holdings. It is ironic that in 1998 the Newhouses despaired of making the publishing profits that they had anticipated. Faced with write-offs in 1997 of $8 million and profits of less than 1 percent, they decided to bail out and sold the firm to Bertelsman, the German publishing empire. What happened in this case illustrates a general trend wherein what were once mostly family businesses became large corporate holdings, which in turn became but small parts of larger groups.

This transformation of the structure of ownership also led to a transformation of the values and expectations governing the business. During the previous decades, book publishers had expected profits of about 4 percent, after taxes. With such returns, many people were not in publishing primarily for the money during those years. Not that Alfred Knopf retired in poverty. He was a wealthy man, and in many respects he deserved to be, as was the case with many other publishers. But it was the slow growth of the firm as a whole that came first, not the annual profit.

As the large conglomerates took over publishing, they assumed that each of their holdings ought to deliver roughly equivalent rates of profit. The logic is simple: if you own cable television stations, lumber mills, and so on, and you also happen to own a book publisher, you would see no reason for the book publisher to make any less than your other holdings. So the pressure is on publishers now to make 10 percent, 12 percent, even 20 percent profits—basically, to quintuple the money they were making before. It's very hard to do that and still publish the wide range of books that had characterized American publishing over the previous decades.

The financial people now running the firms have found structural ways to bring editors—a lot of whom may not feel particularly happy about it—into line with these bottom-line expectations. Each part of the publishing company is broken out into a separate "profit center." Pantheon, for instance, had a very profitable children's line, which effectively subsidized the more demanding—and less immediately profitable—adult books. But once the children's department was accounted for separately, that was no longer possible. Later, the textbook department was separated out, so if a book was sold for course adoption, that money was no longer considered part of the income brought in by the trade department that signed it up. The same thing happened with paperback sales.

The next step was to rationalize profits on a title-by-title basis. It used to be a given in publishing that the best-seller subsidized the other books. But gradually, the pressure was put on to have each book pay its own way. This came to mean not only covering its own costs but also earning enough to cover its share of the overhead. "Small" books, by definition, became impossible to publish. By this reasoning, most large commercial houses in New York now say that if a book can't be expected to sell 15,000 to 20,000 copies in its first year, it shouldn't be published. To get the financial people to agree that something will sell in those numbers, you pretty much have to be able to tell them that the author's previous book has sold that many copies or more. So serious work that may take time to find its audience, whether in the classroom or in paperback, and ambitious work by new authors, become harder and harder to publish. The new Kitty Kelly may receive a $10 million advance, but someone who could be the next Michel Foucault or T. S. Eliot will rarely seem worth the risk.

It isn't that one day the captains of industry came down and said to everyone, "You shall no longer publish what you think ought to be published" or "You may no longer be daring or experimental or avant-garde." Instead, a logical system was gradually imposed and accepted in the industry as a whole, and it gradually became a kind of iron mask that allowed for very little variation. Many saw what was happening and didn't like it. But as everyone knows, most people aren't going to leave their jobs except in extreme circumstances. And because the changes occurred bit by bit over the years, most people in publishing have adjusted to them, and younger people in the business may not even imagine that it has ever been otherwise.

Some publishing houses resisted. Those that were privately owned, such as New Directions and W. W. Norton, continued very much as they had always done, maintaining the very high standards of publishing that they and others had represented. But the firms that were part of larger conglomerate groups had a very different fate. As it happened, it was at Pantheon, at that time under my directorship, that the new pattern was first set.

By 1990, Pantheon had been part of the Random House group for close to thirty years. It was known for its translations of foreign fiction, its innovative social science publishing, and its willingness to publish books in politics and the arts that challenged popular views. For the most part, its books did not make money during their first year. But in the long run, many of the so-called difficult books—such as the works of Noam Chomsky, Foucault, Edward P. Thompson, Eric Hobsbawm, and many others—became staples of a changing curriculum. On a year-by-year basis, a relatively small number of best-sellers brought the firm its major profits. Pantheon was an obvious target for the Newhouse managers, who were intent on imposing the new standards of profitability at all costs.

Robert Bernstein was the head of Random House at the time. For the many years of his tenure, he had adhered courageously to the old standards. Clearly, he stood in the way of the new regime, and in 1990 he was fired from his post. Shortly thereafter, his successor, Alberto Vitale, approached Pantheon with a very simple scheme. Since only a relatively small number of books could be deemed ensured best-sellers, why not cut Pantheon's list—and staff—by two-thirds? This would have the advantage of eliminating all those difficult and troublesome intellectual books (not to mention many of the titles whose content irked Newhouse's political predilections).

Most people in the industry assumed we would accept the offer. Why would we jeopardize our seniority, our comfortable corner offices, the knowledge that our own salaries would be ensured, whatever happened? But my colleagues and I felt as if we were being asked to spend the rest of our publishing careers destroying what we had built, in my case, over a span of thirty years. To everyone's surprise and consternation, all eight of the Pantheon editors, including some very junior ones as well as me, chose to leave rather than give in. At the time, many in publishing thought this an excessive reaction. Surely these were negotiating tactics. People were certain that no publisher would really want to eliminate all vestiges of serious publishing from its list.

One has only to see what has happened, not only to Random House but also to the other conglomerates in the years that have passed since those tumultuous events, to see how much things have changed. Physically, books look better than ever. Overwhelmingly, it is their content that has been transformed. Publishers now see themselves as purveyors of entertainment or information rather than of cultural inquiry. Translations, whether of fiction or nonfiction, have practically disappeared. New and challenging books by younger American scholars likewise are increasingly hard to find. Firms such as Knopf, which once prided

themselves on their broad lists of art criticism, intellectual history, philosophy, and such, now have to admit to authors that they
"can no longer afford" to publish in these areas. The same is true
at the once distinguished firm of Harper & Row (now
HarperCollins) and many others. Indeed, in late 1997, the last
few imprints within the conglomerates devoted to more serious
books were all sold or redirected. HarperCollins sold Basic
Books, even though it was still profitable. The Free Press and
Times Books, both of which published books representing a range
of political opinions, are to be transformed into publishers primarily of business and practical books. Book review editors
lament that they can read through the entire catalog of a major
publisher and find little that is worthy of review.

Increasingly, to find a serious and critical book on contemporary politics, one must turn to the not-for-profit publishers—the Brookings Institution, the Century Foundation (formerly the Twentieth Century Fund), and, I must say, my own
company, the New Press. In the same way, demanding fiction,
literary translations, and poetry are now coming from the new,
independent alternative presses, a wide range of distinguished
houses whose names are less familiar to the American public
but whose catalogs are slowly filling in some of the gaps left
by the retreat of the established houses from these areas. Most
of these, such as the Dalkey Archive Press, Copper Canyon
Press, and Graywolf Press, are located outside New York, and
many are on college campuses, where they can count on inexpensive editorial help and the support of enthusiastic faculty
members.

Can this new array of small firms effectively play the role the
traditional houses once did? My own belief was that this form
presented the only remaining hope for us. Accordingly, after leaving Pantheon, I helped to start a not-for-profit public-interest
publishing firm called the New Press, which in its first five years
published some two hundred books. The publisher of this volume, Island Press, is a similar, though longer established, house.
A look in the Sunday book reviews in newspapers throughout the
country or in the better bookstores will confirm that many of the

books that now challenge the reader are from these new alternatives presses.

But as important as these efforts are, it would be foolish to assume that the culture as a whole has not suffered from the change I describe. These new publishers are all small, usually underfunded, and unable to spend the large amounts needed to commission new work or to get the kind of distribution and store placement the large firms are able to command. Along with the university presses, they represent much less than 1 percent of total book sales in the country.

In this Puritan-influenced culture, the concept of the saving remnant still plays an important role. To many, the fact that these books still appear, that new firms have been created, is a sign that ours is a basically healthy society, one capable of developing new forms to supplant the old. There is something to be said for this optimistic reading. Particularly in fiction and poetry, it is probably realistic to say that a very good book has as good a chance of being published as it would have had some twenty or thirty years ago. In nonfiction, however, the chances are far different. Books that require a substantial advance to enable their authors to spend several years on research are less and less likely to be funded.

The decision to take on a book or a book proposal is now basically made according to the simplest of criteria: will the idea sell? Not is it interesting or important but is it hot, commercial, popular? The plight of serious nonfiction shows dramatically the degree to which ideas have become commodities whose value can be measured by the number of potential customers. By definition, dissenting or countercyclical ideas are far less likely to find a publisher. A form of market censorship has firmly established itself, the results of which are apparent in our bookstores. True, an important number of books funded by right-wing foundations have been an exception to this rule. The right, wisely, does not entrust its future to the marketplace. But books that challenge the status quo from other vantage points are far rarer. It used to be an American tradition that each election year would see a spate of books discussing the major issues of the day. One has only to look at the questions we face as a country now, whether the future of the welfare state, the need for more international trade agreements, or the case for intervention overseas, to realize how few books on these issues were published in 1996 or, indeed, in 1992. The careful thought and argument needed for such discussion were not in evidence on the lists of the major publishers.

Nor can we count on newspapers and magazines to commission this kind of research. The same emphasis on profit has been even stronger in the mass media than in book publishing. This is a subject for another essay in itself, but suffice it to say that newspapers are less likely to send reporters on ambitious and time-consuming assignments, and television and radio news programs are even less so. The days when the National Broadcasting Corporation paid for a full-time symphony orchestra under the leadership of Arturo Toscanini are well behind us.

How, then, will American society debate the issues of the future? How will we know of new ideas, different approaches, countercyclical and dissenting ideas? That is the crucial question posed by the transformation of book publishing. There may well be new Edward Bellamys and Henry Georges and Harriet Beecher Stowes ready to capture our imaginations or pose new solutions to the problems we face. In the past, publishing played an honorable role—indeed, a crucial one—in disseminating new ideas to the broadest of publics. Whether this will continue to happen in the next century, when America will be increasingly challenged, is a question that has to be left open.

Movies and the
Selling of Desire

Molly Haskell

BACK IN THE 1960s and 1970s, fresh-faced Americans fleeing to Paris to escape the materialism back home would emerge from Left Bank movie theaters dumbfounded by the crass commercials. How could the French, those high priests of culture, those royalists of taste, be guilty of such blatant commercialism when our own arriviste country, the cradle of advertising, kept its cinemas free of the huckster's pitch!

Hollywood and Madison Avenue, though they might be sleeping together, still maintained a formal separation, and if one paid the price of admission, one was not expected to endure a sponsor's blandishments. New York audiences, myself among them, would hurl boos and catcalls when the Coca-Cola Company or the American Express Company tried to sneak in a plug. Thus chastened, the "sponsor" would skulk away and then try again a few years later, only to elicit the same lynch-mentality response. That was then. Now, viewers sit in numbed (or, some maintain, even delighted) silence, bombarded by one commercial after another as the feature is endlessly postponed. As often as not, when the feature finally arrives, it not only is filled with "product placements," thus serving as an extension of the commercial pitch, but also is directed by veterans of MTV or television commercials with the same blitzkrieg style of quick cutting, packed images, and a high-voltage rhythm of carefully timed climaxes. For those purists who would prefer to absent themselves during the commercials (as is still an option for French audiences, where theatres sequester the feature film from *le cinéma publicité*, commercials are cunningly sandwiched between coming attractions: one trailer, one commercial; one trailer, one commercial. Adding to the blather are those tired pleas, also known as public service announcements, instructing the amorphous masses how to comport themselves in the temple of cinema, which is already reeking with the smell of stale popcorn oil and crackling with the sound of munched refreshments. Moreover, the trailer itself is no longer a soft-sell teaser for things to come; it is an all-out assault, a miniature version of the movie, complete with every climax and plot development, jammed into a manic, take-no-prisoners pitch. It pops up on television, the movie's adjunct in promotion, along with ancillary blanket-the-media "profiles" and star plugs to hook that crucial teenage audience, which will rush to the box office on that all-important opening weekend.

Remember the outcry that greeted subliminal advertising in the 1950s? Clergymen, pundits, and other spokesmen for the body politic cried foul at the prospect of advertisements arriving on padded feet (we should be so lucky) to snatch our souls.[1] We're told that today's young people have been raised in an atmosphere so suffused with advertising that it's part of the visual and aural wallpaper, nothing to get excited about. Media frenzy and the cacophony of constant hyperbole have spawned mutants in tune with their own rhythms, so that those who talk about the blurring of the line between advertising and content are speaking in anachronisms, advertising their own obsolescence. The "youth" audience born of the 1960s counterculture, now sought, coveted, and emulated like a sacred icon, has become so embedded as both means and end in the thinking of moviemakers and advertisers that it has acquired the aura of an absolute, to be feared and sought after but never questioned. In the demographic terms that now function as holy writ, the youth market consists of not only American teenagers but also a global audience of action movie fans, which is to say a lowest-common-denominator mass of minimal literacy and maximal impressionability and buying power.

But as consumers of film, and as film-made consumers, are they a different breed or merely a slightly more infantilized and product-friendly version of . . . previous generations? And how justified are we who lament these processes in proclaiming the end of the civilized world as we know it, when so many good movies are still being churned up and onto the beach from what French New Wave director Claude Chabrol once referred to as the ocean of cinema? Aren't movies, after all, in the business of seduction, and even before Hollywood became so overtly mercenary and relentlessly bottom-line oriented, weren't movies softening us for the kill?

So, before succumbing to Cassandra-like gloom or warbling over the familiar decline-and-fall scenario, the question to ask is, How good were the good old days, and how pure were the movies of the once despised, now lamented studio system? Does the quantification of everything into attendance figures and profit motives (in museum attendance and book buying as well as in more popular forms of entertainment) represent a change in substance or merely in scale?

Movies, American division, were born and delivered not into a pristine laboratory of pure art or pure technology but into an urban demimonde, raffish and opportunistic. What would become the cinema had no sooner emerged from the toy-and-experiment stage than it was enveloped in a flurry of promotional zeal by its early champions—Thomas Edison in the United States and the Lumière brothers in France. Just as early movies were one or two steps from the peep shows, the Jews who found easy entry into the upstart industry and became its moguls were one or two removes from the shtetl, and achieved their success by appealing to a similarly disenfranchised audience of immigrants. For these spectators, generally impoverished and non–English-speaking two-reelers were both the only affordable entertainment and an instant lesson in the customs and behavior of the Wasp elite as imagined by Hollywood. One could say that movies have in some sense come full circle to their illiterate roots, now dumbing down for primitive audiences where they once smartened up.

Studio filmmaking was by definition authoritarian, a hegemony run by a group of autocratic arrivistes who bolstered social insecurity with an exaggerated sense of moral responsibility. Like the bastard child it was, born of art and commerce, sleaze and glamour, the medium was always in danger of losing its thin veneer of respectability and slipping into perdition. In the 1920s and early 1930s, Hollywood was America's own Sodom and Gomorrah. Rushing to put out the fires of scandal, Hollywood instituted the self-policing Production Code, whose quaintly worded dos and don'ts perfectly captured that disarming mixture of public-spirited rhetoric and face-saving hypocrisy. It was not the first or the last time we will find in movies a crossroads where a vestigial world of Victorian morals meets the amoral world of capitalist opportunism.

From their inception, movies—both pusher and product—fed, paralleled, and to some extent created consumerism. The growth of the moving pictures before and after World War I came at a time of intense industrial growth in general and a proliferation of new products. It was a high-flying, money-making era into which advertising inserted itself as a means of exploiting new markets and selling surplus goods. Advertising was quickly transformed from a mechanism that united supply and demand to a perpetual motion machine that pursued customers and maintained objectless appetites, especially when times were lean and buying power had shrunk. Hand in hand with the role of movies in the emerging drama of consumerism was the rise of the department store at the end of the nineteenth century. In the pioneering art of display, we see the triumph of packaging over substance as moviemakers and purveyors of goods, like expert politicians, instinctively home in on unconscious needs, cravings that are inchoate and not yet articulated yet are rising, rising to meet the bait. The same aesthetic can be seen in the emergence and subsequent packaging of Hollywood stars. These fabulous individuals, soon to demand close-up shots and backlighting, were initially created not by the parsimonious producers, who would have preferred to keep actors poor and anonymous, but by movie viewers who began recognizing certain performers from picture to picture, preferring one to another and demanding more "inside" information about their favorites.

Consumerism and the appetite for movies were in many ways *the* binding force of an America composed of disparate racial and religious groups, languages, and economic strata. In a sharply divided, class-bound society, movies offered a vision of homogeneity and access to the better things of life, in which everyone could share. The great irony is that what began as a unifying influence, integrating family values with an ideal of mainstream, middle-class culture, eventually became a source of division and fragmentation as movies—in the endlessly adaptive, democratizing pattern of capitalism itself—mutated into avatars of movement and change, brandishing freedom, flight, and a restless individualism.

In the history of movies, one can see the transition from the enshrinement of Victorian virtues such as hearth, sacrifice, and duty to an embrace of the more sybaritic twentieth-century ideals of self-interest and fulfillment. In her provocative book *The Demoralization of Society,* historian Gertrude Himmelfarb makes the point that contrary to much twentieth-century stereotyping, those Victorian virtues weren't a set of principles by which a stuffy and narrow-minded bourgeoisie held in check its social inferiors. The "virtues"—a stern set of ideals, as distinct from today's flabbier "values"—were in fact the social glue, with a premium on respectability as the great quality, that united the lower and middle classes.[2]

With the emerging American middle class, respectability and

consumerism became curiously enmeshed. Because of movies—the visual medium most instrumental in the twentieth-century ascendancy of image—the notion of appearances, once linked to notions of self-respect, gradually came to have less to do with character and reputation and more to do with simply looking good.

In the emphasis on the visual, in the conditioning of our eyes toward the project of consumption, looking and buying become intertwined. Movies are the ideal arena for gazing, yearningly, at images that intoxicate and soothe, inspire and excite, stimulating the passive enthrallment that nurtures vague but powerful longings.

In contrast to the current marketing strategy, which is directed almost exclusively at young males, the ideal viewer and consumer of the cinema of earlier decades was female. Whereas in the biblical Genesis the first occupant of the Garden of Eden was a man, in the consumer paradise launched by movies in the twentieth century the first consumer was a woman—she was the one who bought and the one who imagined herself arrayed in the raiments and aura on display. In the 1930s, she might have been someone like Cecilia, the character portrayed by Mia Farrow in Woody Allen's *The Purple Rose of Cairo,* a low-paid working girl or a melancholy housewife who returns again and again to lose herself—and escape the woes of the Great Depression—in the spectacle of Park Avenue swells tripping the light fantastic and whispering words of love. In the1920s and 1930s—in fact, in every moviegoing decade until the 1970s—women were the primary audience for movies in the daytime and, when couples went together, the ones who decided what movie to see. In contrast to the actual position of women in society at the time, actresses in Hollywood (in movies often written by female screenwriters) made as many movies as their male counterparts and had equal billing and equal—or sometimes higher—salaries.

In both Olive Higgins Prouty's 1923 novel *Stella Dallas* and its movie adaptations (particularly the classic 1937 tearjerker directed by King Vidor and starring Barbara Stanwyck), the ambitious but hopelessly déclassé heroine reads fan magazines and goes to movies to both satisfy and give substance to her romantic yearnings.[3] Her dreams of a better life, linked to a dashing male rescuer, are vague and urgent. Like some backwater Emma Bovary, she is led by the incantatory power of religious and pseudoreligious imagery into a sort of narcissistic trance in which the material and the spiritual merge in a cult of self-improvement and social betterment through love.

Much theoretical writing on the "woman spectator" of the past is implicitly condescending, portraying her as a victim of propaganda and a pliant object of Pavlovian conditioning. However susceptible women were to the wares on display, I'd suggest that what the female viewer is buying, or buying into, is not just the accoutrements of the "good life"—in *The Purple Rose of Cairo*, the Park Avenue penthouse, the champagne, the sophistication (the "madcap Manhattan weekend," as Jeff Daniels's dashing, ingenuous explorer cries)—but also another version of herself. That ideal self is not only stylishly clothed and housed but also possessed of higher sentiments and deeper thoughts. For in the peculiar dynamic of identification with film stars, a kind of transubstantiation occurs. In those luminous close-ups, there is little separation between the material and the spiritual, face and soul, and things and feelings, the former being the outer envelope for the latter. The beauty of those iconic faces filling the screen becomes a metaphor for inner beauty.

For both Stella (Barbara Stanwyck), stuck in her milltown, and Cecilia (Mia Farrow), trapped in an abusive marriage, the shabbiness of life becomes intolerable. For both women, the focus of yearning is a not-entirely-fictional world that by its very existence gives shape and a voice to their discontent. Movies have so often been condemned as an opiate, an escape hatch that allows the miserable to return obediently to their unhappiness, that their revolutionary power in providing an outlet and a language for aspiration is frequently overlooked. Marxists and sociologists who are presumably immune to their power tend to see movies purely as a tool of capitalism that acts in various nefarious ways—as a seductive shill for materialism, a palliative that ensures resignation, or an instrument for stirring up a discontent that has nowhere to go. In fact, movies have played a role both progressive and conservative, democratizing in their populist appeal, conformist in their acceptance of the status quo, and paradoxical in the built-in, ongoing conflict between elitism (the stars) and egalitarianism (the story).

Audiences wept (and still weep) with Stella as she gives up her daughter, Laurel, the love of her life, to the well-placed Helen Morrison. The socially prominent and now widowed old flame of Laurel's father can offer "more" in the way of clothes, education, and, above all, marriage to an upper-class beau. But no one sees this as an absolute wrong, so essential to 1930s thinking is the fantasy of betterment and so hopelessly short of the ideal is Stella's way of life. Sunk to the bottom rung of the social ladder, raucous in her choice of friends and furbelows, she struggles to cling to respectability, but her efforts are compromised by her native flamboyance.

Learning from the movies how to look and how to dress— acquiring the refinements of class—was a worthy goal, but it was one to which the irrepressibly tacky Stella could only aspire vicariously. Her daughter, however, born with better opportunity and instruction and a working quantity of upper-class genes inherited from her father, would slip easily into the genteel preserves of upper-crust privilege—though not without the anguish and guilt of betrayal.

Women's primary vocation, and means to the end of social ascent, was to make themselves—ourselves—into objects of desire. The Stellas and Cecilias, the immigrants and the working-class poor, were aspirants to a middle class in which romance governed matrimonial choice. Movies and the love songs often inserted into them would teach us how courtship worked: what to say; how to dress, kiss, make love, and break up; and, above all, the nuances of class, the barriers that might be crossed and toppled. Movies of the 1920s and 1930s gave women lessons in, say, monogamy, or in how to deal with a straying husband, but the message was becoming more radical—from avert your eyes and smile bravely in the 1920s to avert your eyes *or* go to Reno and get divorced in the 1930s. The movie version of Clare Booth

Luce's play *The Women* features samples of each: Norma ("Chin-Up") Shearer as the bravely enduring wife and sassy Paulette Goddard as the free-spirited Reno graduate.

The common wisdom among feminist theoreticians is that the process of identifying with the desirable film star involved a woman's turning herself into a commodity, an object of the male "gaze." A version of "Men act, women react," women's position in the agenda of sexuality and "difference" as defined by the psychoanalytic gurus Sigmund Freud and Jacques Lacan is not that of agents, doers, but of the Other, fetishized, distanced, and neutered by male fears and fantasies. But when women are cast as the doomed ladies-in-waiting, what is rarely acknowledged is the extent to which the strong female presences change and complicate the terms of such fixed binary formulations, exerting power far beyond the literal outlines of their roles.

Movies of the 1930s, awash in class conflict, upward striving, and therefore hope, featured women who, like the heroines of Jane Austen's emerging middle class, might actually choose a husband rather than have him chosen for her: consumers, then, not only of hats and haberdashery but also of matrimonial mates. Women, empowered by suffrage, were moving further toward sexual equality, which was envisioned and alluded to but whose difficulties were foretold in stories that lifted a woman out of the ordinary only to send her scampering back into marriage.

Movies have always been in the business of selling themselves along with whatever products might be linked with them through merchandising tie-ins. Longing was itself a story of ambiguity, of hopes raised and frustrated, of desire endlessly prolonged and never satisfied.

Freud described sexual desire as doomed to frustration because something in its very nature was inimical to fulfillment. This isn't so mysterious if one considers that the source of that desire is the child's longing for undifferentiated union with the mother, a longing that comes into being only when separation has occurred and the sense of an individual "I" is born. This paradox is duplicated in our infatuation with the stars—indeed, it is at the heart of their magic, a hold over us that is based on a sense of in-

timacy and merging that at the same time is pure illusion. The
stars are both our servants and our masters. We pay our money
and accept or reject them in a way that we couldn't accept or reject our parents: being our "'property," at least momentarily, they
are our revenge for the humiliation of childhood and its many im-
potencies. But then, being owned by everyone, they are owned by
no one; hence our resentment, our fickleness, our satisfaction in
the gossip retailing their misfortunes. They resent us for encroaching on their lives, and we resent them for being so much
more lustrous and powerful than we are and for exposing our
neediness; and so, like the Greeks and their beautiful, capricious
gods, we go on needing and feeding on each other. As we come to
understand the world of brain chemistry and neurotransmitters,
of pheromones and dopamine, of the way addiction works—the
buzz of shopping, the endorphin rush of getting and spending, of
watching and wanting—we may come to realize that movies have
permanently altered the hardwiring of our response mechanisms.
We eagerly consume every chapter, verse, and document of a
presidential scandal—what movie can equal its combination of
star power and real-life danger?—but it plays this way to a recep-
tive audience because we have been prepared for both its form
and its content, its once unmentionable language, by decades of
vicariously scandalous scenarios in the movies.

As Alfred Hitchcock slyly gave us to understand in film after
film, we are all voyeurs, living through our stars, implicated in the
nasty business of an idolization fraught with self-loathing, sadism,
and betrayal. Movies, and now television, are our *Rear Windows*;
from the safety of our living rooms we watch while our dazzling
surrogates walk into the lion's den on our behalf.

Movies are and always have been a revolutionary force. Like
capitalism itself, they seek out new markets, acknowledge no national boundaries, and, in their narratives, promote a materialistic way of life. By virtue of their kinetic nature, they make ever
more modern forms of transportation fashionable and desirable.
Because physical mobility is at a premium today, we no longer
have sagas of generations of a family living in a single place; instead, we have fast cuts between one city or country and another,

across boundaries of time and place. A friend, a baby-boomer dad, described taking his twelve year-old son to see the latest James Bond movie. The boy was enthralled, but the father, exasperated by the movie's brain-dead freneticism, could barely sit still. He brought his son home and rented a video of one of the Sean Connery "classics" he'd been brought up on. This time, it was the boy who fled in boredom, leaving the dad to contemplate the wasteland of today's adolescent tastes compared with the golden age of *Goldfinger* and *Dr. No*. The James Bond films are a virtual time capsule of the artifacts and attitudes of late-twentieth-century consumerism: the minefield of disposable gadgetry; the processing of global events and alliances in the Cuisinart of throwaway plots; continents and vehicles consumed by the motor of perpetual wanderlust; and the latest models of disposable babes and gun-toting bimbos, with the most recent Bond movie proudly advertising a more "active" female protagonist, Asian to boot, from whom we are unlikely to hear further. Movies both reflect and create social conditions, but their special charm is to offer fantasy clothed as virtual reality, a world where people consume without the tedium of labor. Characters arrive with resources and possessions for which no explanation is necessary, no acknowledgment of the effort that went into acquiring them. They float in a credit card world where transactions are never seen and the bill never comes due . . . and we wonder why we're a debtor nation!

With their ever fresh supply of new stars and new faces to breathe life into even the hoariest and most overused of plots, movies are flagships in the armada of novelty-seeking conquistadores, expensive toys unabashedly brandishing populist credentials. The $200 million *Titanic,* the most expensive movie ever made, celebrates the proletarian "real folk" in steerage over the nasty swine in first class. It seems that to give ideological credibility to the business of consumerism, every generation and every decade must keep afloat a pose of rebellion: youth against old age, hip against square, the creative and transgressive as opposed to the stuffy and narrow.

In his stimulating book *The Conquest of Cool,* Thomas Frank challenges the typical 1960s view of corporate culture as a stalwart behemoth.[4] He points out that big companies as led by the trendoids and anarchists of Madison Avenue were just as ready to break out of the mold as any peace marcher, bra burner, or tie-dye-wearing flower child. Moreover, far from merely co-opting the rhetoric of rebellion—as in the Pepsi and Volkswagen campaigns that promoted youth and fun and attacked the hypocrisies of advertising, for example—Madison Avenue was in fact undergoing a parallel revolution of its own—against dullness and hierarchy, against gray-flannel-suit conformity—that found in the language of the counterculture its own revolutionary voice.

Fast-forward to the millennial 1990s, and what is the symbol that unites the world at the Olympics campfire? Not the interlocking circles of international harmony but the Nike Swoosh, symbol of speed and international commerce. Kids in poor neighborhoods steal for it; television reporters compromise their neutrality by sporting the logo on their windbreakers; William Burroughs trumpets them over songs by the Beatles.

The pervasive emphasis on what's hot and what's not makes for an atmosphere frantic with anxiety. Will the Spice Girls be "out" before we can find out who they really are, and is it worth the trouble? The sheer quantity of movies—some four hundred per year in the late 1990s, as opposed to only half that in the early part of the decade—makes it difficult to reflect in leisure or to hold on to images of singularity while we are prodded by editorial pressure to leave last week's work of art behind and skip to the beat of the New and the Now. Of course there is nothing new in the pressure to conform to the tempo of whatever "now" one happens to be living in. Sixty years ago, movies gleefully recorded the battle of the older generation to resist the blandishments of jazz and swing in popular music. At the same time, many movies waxed nostalgic over the good old songs all the way back to the Gay Nineties. Thus, the more leisurely tempos of the past coexisted with the speeded-up pace of modernity, just as today a literary culture, however beleaguered, coexists with (and occasionally overlaps and exploits) an Internet culture. Furthermore, despite all the mind-altering and ever more obtrusive devices to condition us to the rhythms and religion of consumerism, there remains for some of us a precious margin of dissent, irreverence, and histori-cal perspective. For example, one trend working against the pre-sumed disposability of movies and other artifacts of popular cul-ture is the growth of preservation initiatives that enable us to see that the way we are is not fundamentally different from the way we were.

The Ecology of Giving and Consuming

David W. Orr

What one person has, another cannot have . . . every atom of substance, of whatever kind, used or consumed, is so much human life spent.

—John Ruskin, *Unto These Last*

How do we sell more stuff to more people in more places?

—IBM advertisement

Don't try to eat more than you can lift.

—Miss Piggy

SOME YEARS AGO a friend of mine, Stuart Mace, gave me a letter opener he had carved by hand from a piece of rosewood. Over his seventy-some years Stuart had become an accomplished wood craftsman, photographer, dog trainer, gourmet cook, teacher, raconteur, skier, naturalist, and all-around legend in his hometown of Aspen, Colorado. In the mountains above Aspen, Stuart and his wife, Isabel, operated a shop called Toklat (in the Eskimo language, the term means "alpine headwaters"), featuring an array of woodcrafts, Navajo rugs, jewelry, fish fossils, and photography. He used his "free" time in summers to rebuild parts of a ghost

137

town called Ashcroft for the USDA Forest Service. He charged
nothing for his time and labor. For groups venturing up the mountain from Aspen, he and Isabel cooked dinners featuring local
foods prepared with style and simmered over great stories about
the mountains, the town, and their lives. Stuart was seldom at a
loss for words. His living, if that is an appropriate word for the
way a Renaissance man earns his keep, was made as a wood-
worker. He and his sons crafted tables and cabinetwork with ex-
quisite inlaid patterns using an assortment of woods from forests
all over the world. A Mace table was like no other that I've seen.
Long before it was de rigueur to do so, Stuart bought his wood
from forests managed for long-term ecological health. The cali-
bration between ecological talking and doing wasn't a theoretical
thing for Stuart. He paid attention to details.

I first met Stuart in 1981. I was living in the Ozarks at the time
and was part of an educational organization that included, among
other things, a farm and a steam-powered sawmill. In the summer
of 1981 one of our projects was to provide two tractor-trailer loads
of oak beams for the Rocky Mountain Institute, which was being
built near Old Snowmass, Colorado. Stuart advised us in the
cutting and handling of large timber, about which we knew
little. From that time forward Stuart and I saw each other several
times a year, either when he traveled through Arkansas or when
I wandered into Aspen in search of relief from Arkansas summers. He taught me a great deal, not so much about wood per
se as about the relation of ecology, economics, craft work, generosity, and good-heartedness. I last saw Stuart in a hospital room
shortly before he died of cancer, in June 1993. In that final conversation, I recall Stuart being considerably less interested in the
cancer that was consuming his body than in the behavior of
the birds outside his window. He proceeded to deliver an im-
promptu lecture on the ecology of the Rocky Mountains. We
cried a bit and hugged, and I went on my way. Shortly thereafter
he went on his.

Every time I use his letter opener, I think of Stuart. I believe
that he intended it to be this way. For me the object itself is a les-

son in giving and appropriate materialism. First, it is a useful thing. Hardly a day passes that I do not use it to open my mail, to pry something open, or as a conversational aid to help emphasize a point. Second, it is beautiful. The coloring ranges from a deep brown to a tawny yellow. The wood is hard enough that it does not show much wear after a decade and a half of continuous use. Third, it was made with great skill and design intelligence. The handle is carved to fit a right hand. Two fingers fit into a slight depression carved in the base. My thumb fits into another depression along the top of the shank. It is a pleasure to hold; its smoothness feels good to the touch. And it works as intended. The blade is curved slightly to the right, which serves to pull the envelope open as the blade slices through the paper.

Had Stuart been a typical consumer he could have saved himself some time and effort. He could have hurried to a discount office supply store to buy one of the cheap chrome-plated metal letter openers stamped out by the tens of thousands in some "developing" country by underpaid and overworked laborers employed by a multinational corporation using materials carelessly ripped from the earth by another footloose conglomerate and shipped across the ocean in a freighter spewing Saudi crude every which way and sold by nameless employees to anonymous consumers in a shopping mall built on what was once prime farmland and is now uglier than sin itself, making a few dollars for some organization that buys influence in Washington and seduces the public on television. But you get the point.

In other words, had Stuart been a "rational" economic actor he would have saved himself a lot of time, which he could have used for watching the Home Shopping Channel. He could have maximized his gains and minimized his losses, as the textbooks say he should do. Had he done so, he would have been participating in that great scam called the global economy, which means helping some other country "develop" by selling the dignity of its people and their natural heritage for the benefit of others who lack for nothing. And he would have helped his own country's gross national product become all that much grosser.

A GREAT GLOBAL DEBATE is under way about the sustainability
and fairness of present patterns of consumption.[1] On one side are
those speaking for the poor of the world, various religious organizations, and the environment and argue adamantly that wealthy
Americans, Japanese, and Europeans consume far too much.
Doing so, they believe, is unfair to the poor, to future generations,
and to other forms of life. It is stressing the earth to the breaking
point. Others, who believe themselves to be in the middle, argue
that we do not consume too much; rather, we consume with too
little efficiency. Below the surface of such views there is, I suspect, the gloomy conviction that it is too late to rein in the hedo-
nism loosed on the world by the advertisers and the corporate
purveyors of fun and convenience. Human nature, they think, is
inherently porcine, and given a choice, people wish only to see
the world as an object to consume, with the highest purpose of
life to maximize bodily and psychological pleasure. For the managers, a better sort, a dose of more advanced technology and better organization will keep the goods coming. No problem. This
view of human nature I take to be a self-fulfilling prophecy of the
kind Fyodor Dostoyevsky's Grand Inquisitor would have appreciated. At the other end of the debate are the economic buccaneers
and their sidekicks who talk glibly about more economic growth
and global markets. A quick review of the seven deadly sins reveals them to be full-fledged heathens who will burn for eternity
in hellfire. (I am the son of a Presbyterian preacher.)

Because I believe that it is right and because I know it needs
help, the first position in this debate is the one for which I intend
to speak. I must begin by noting that to consume, as defined by
the *Oxford English Dictionary,* means to "destroy by or like fire or
formerly disease." A "consumer," then, is "a person who squanders, destroys, or uses up." In this older and clearer view, consumption implied disorder, disease, and death. In our time, however, we proudly define ourselves not so much as citizens or
producers or even as persons, but as consumers. We militantly
defend our rights as consumers while letting our rights as citizens
wither. Consumption is built into virtually everything we do. We
have erected an economy, a society, and what will soon be an en-

tire world around what was once recognized as a form of mental derangement. How could this have happened?

The emergence of the consumer society was neither inevitable nor accidental. Rather, it resulted from the convergence of four forces: a body of ideas saying that the earth is ours for the taking; the rise of modern capitalism; technological cleverness; and the extraordinary bounty of North America, where the model of mass consumption first took root. More directly, our consumptive behavior is the result of seductive advertising, entrapment by easy credit, prices that do not tell the truth about the full costs of what we consume, ignorance about the hazardous content of much of what we consume, the breakdown of community, a disregard for the future, political corruption, and the atrophy of alternative means by which we might provision ourselves. The consumer society, furthermore, required that human contact with nature, once direct, frequent, and intense, be mediated by technology and organization. In large numbers we moved indoors. A more contrived and controlled landscape replaced one that had been far less contrived and controllable. Wild animals, once regarded as teachers and companions, were increasingly replaced with animals bred for docility and dependence. Our sense of reality, once shaped by our complex sensory interplay with seasons, sky, forest, wildlife, savanna, desert, river, sea, and night sky, increasingly came to be shaped by technology and artificial realities. Urban blight, sprawl, disorder, and ugliness have become, all too often, the norm. Compulsive consumption, perhaps a form of grieving or perhaps evidence of mere boredom, is a response to the fact that we find ourselves exiles and strangers in a diminished world that we once called home.

Since stupidity is usually a sufficient explanation for what goes wrong in human affairs, theories of conspiracies that require great cleverness are generally both superfluous and improbable. In this case, however, there is good reason to think that both were operative. Clearly, we were naive enough to be suckered by folks such as department store magnate Lincoln Filene and General Motors executive Alfred Slone, who conspired to create the kind of human beings who could be dependably exploited and could even

come to take a perverse pride in their servitude. The story has
been told well by Thorstein Veblen, Stuart Ewen, William Leach,
and others and does not need to be repeated in great detail here.[2]
In essence, it is a simple story. The first step involved bamboo-
zling people into believing that who they were and what they
owned were one and the same. The second step was to deprive
people of alternative and often cooperative means by which they
might fulfill basic needs and obtain basic services. The destruction of light-rail systems throughout the United States by General
Motors Corporation and its coconspirators, for example, had
nothing to do with markets or public choices and everything to do
with backroom deals designed to destroy competition for the automobile. The third step was to make as many people as possible
compulsive and impulsive consumers, which is to say addicts, by
means of daily bombardment with advertising. The fourth step required that the whole system be given legal standing through the
purchase of several generations of politicians and lawyers. The
final step was to get economists to give the benediction by announcing that greed and the pursuit of self-interest were, in fact,
rational. By implication, thrift, concern for others, public mind-
edness, farsightedness, and self-denial were old-fashioned and ir-
rational. Add it all up, and voilà! The consumer: an indoor, plea-
sure-seeking species adapted to artificial light, living on plastic
money, and unable to distinguish the "Real Thing," as in Coca-
Cola, from the real thing.

Do we consume too much? Certainly we do! In businessman
and economist Paul Hawken's words:

> Americans, who have the largest material requirements in
> the world, each directly or indirectly use an average of 125
> pounds of material every day, or about 23 tons per year. . . .
> Americans waste more than 1 million pounds per person per
> year. This includes: 3.5 billion pounds of carpet sent to land-
> fills, 25 billion pounds of carbon dioxide, and 6 billion
> pounds of polystyrene. Domestically, we waste 28 billion
> pounds of food, 300 billion pounds of organic and inorganic
> chemicals used for manufacturing and processing, and 700
> billion pounds of hazardous waste generated by chemical

production. . . . Total wastes, excluding wastewater, exceed 50 trillion pounds a year in the United States. . . . For every 100 pounds of product we manufacture in the United States, we create at least 3,200 pounds of waste. In a decade, we transform 500 trillion pounds of molecules into nonproductive solids, liquids, and gases.[3]

Does compulsive consumption add to the quality of our lives? Beyond some modest level, the answer is no.[4] Does it satisfy our deepest longings? No; nor is it intended to do so. To the contrary, the consumer economy is designed to multiply our dissatisfactions and dependencies. In psychologist Paul Wachtel's words, "Our present stress on growth and productivity is intimately related to the decline in rootedness. Faced with the loneliness and vulnerability that come with deprivation of a securely encompassing community, we have sought to quell the vulnerability through our possessions."[5]

Do we feel guilty about the gluttony, avarice, greed, lust, pride, envy, and sloth that drive our addiction? A few may. But most of us, I suspect, consume mindlessly and then feel burdened by having too much stuff. Our typical response is to hold a garage sale and then take the proceeds to the mall and start all over again. Can the U.S. level of consumption be made sustainable for all 5.8 billion humans now on the earth? Not likely. By one estimate, to do so for just the present world population would require the resources of two additional planets like the earth.[6]

If there ever was a bad deal, this is it. For a mess of pottage we surrendered a large part of our birthright of connectedness to one another and to the places in which we live, along with a sizable part of our practical competence, intelligence, health, community cohesion, peace of mind, and capacity for citizenship and neighborliness. Our children, consumers-in-training, can identify more than a thousand corporate logos but only a dozen or so plants and animals native to their region. As a result they are at risk of living diminished, atomized lives. We consume, mostly in ignorance, chemicals such as atrazine and alachlor in our cornflakes, formaldehyde in our plywood and particle board, and perchloroethylene in our dry-cleaned clothing.[7] Perhaps as many as five hundred other synthetic chemicals are embedded in our fatty tissues, with effects on our health and behavior that we will never fully understand. Our rural landscapes, once full of charm and health, are dying from overdevelopment, landfills, discarded junk, too many highways, too many mines and clear-cuts, and a lack of competent affection. Cities, where the civic arts, citizenship, and civility were born, have been ruined by the automobile. Death by overconsumption has become the demise of choice in the American way of life, with the death certificates listing cancer, obesity, and heart disease. Some of our kids now kill each other over Nike shoes and jackets with NFL logos. Tens of thousands of us die on the highways each year trying to save time by consuming space. To protect our "right" to consume another country's oil we have declared our willingness to incinerate the entire planet. We have, in short, created a culture that consumes everything in its path, including our children's future. The consumer economy is a cheat and a fraud. It does not—indeed, cannot—meet our most fundamental needs for belonging, solace, and authentic meaning.

"WE MUST," writes Wendell Berry, "daily break the body and shed the blood of creation. When we do this knowingly, lovingly, skillfully, reverently, it is a sacrament. When we do it ignorantly, greedily, clumsily, destructively, it is a desecration."[8] Can our use of the world be transformed from desecration to sacrament? Is it possible, in other words, to create a society that lives within its ecological means—taking no more than it needs, replacing what it takes, depleting neither its natural capital nor its people—one that is ecologically sustainable and also humanly sustaining?

The general characteristics of that society are, by now, well known. First, a sustainable society would be powered by current sunlight, not ancient sunshine stored as fossil fuels. Prices in such a society would reflect, as Henry David Thoreau put it, "the amount of life exchanged for a thing," which is to say the full costs of things. It would not merely recycle its waste but would eliminate the very concept of waste. Since "the first law of intelligent tinkering," as Aldo Leopold once wrote, "is to keep all of the pieces," a sustainable society would hedge its bets by protecting both biological and cultural diversity.[9] Such a society would exhibit the logic inherent in what is called system dynamics, having to do with the way things fit together in harmonious patterns over long periods of time. Its laws, institutions, and customs would reflect an awareness of interrelatedness, exponential growth, feedback, time delays, surprise, and counterintuitive outcomes. It would be a smarter, more resilient, and ecologically more adept society than the one in which we now live. It would also be a more materialistic society in the sense that its citizens would value all materials too highly to treat them casually and carelessly. People in such a society would be educated to be more competent in making and repairing things and in growing their food. They would thereby understand the terms by which they are provisioned more fully than most of us do today.

There is no good argument to be made against such a society—all the more reason to wonder why we have been so unimaginative and so begrudgingly slow to act on what later generations will see as merely an obvious convergence of prudent self-interest and ethics. It is certainly not for a lack of spilled ink, conferences in exotic places, and high-powered rhetoric. But for most of us most of the time, sermons aiming to make us feel guilty about our consumption seldom strike a deep enough chord. The reason, I think, has to do with the fact that we are moved to act more often, more consistently, and more profoundly by the experience of beauty in all its forms than by intellectual arguments, abstract appeals to duty, or even fear.

The problem is that we do not often see the true ugliness of the consumer economy and so are not compelled to do much about it. The distance between the shopping malls and their associated mines, wells, corporate farms, factories, toxic dumps, and landfills, sometimes half a world away, dampens our perception that something is fundamentally wrong. Even when visible to the eye, the ugliness of such systems is concealed from our minds by their very complexity, which makes it difficult to discern cause and effect. It is veiled by a fog of abstract numbers that measure our sins in parts per billion and as injustices discounted over decades and centuries. It is cloaked by the ideology of progress that transmutes our most egregious failures into chrome-plated triumphs.

We have models, however, of a more transparent and comely world, beginning with better ways to provide our food, fiber, materials, shelter, energy, and livelihood and to live in our landscapes. Over the past 3.8 billion years, life has been designing strategies, materials, and devices for living on the earth. The result is a catalog of design wisdom vastly superior to the best of the industrial age. This wisdom might instruct us in the creation of farms that function as do prairies and forests, wastewater systems modeled after natural wetlands, buildings that accrue natural capital as trees do, manufacturing systems that mimic ecological processes, technologies with efficiencies that exceed those of our best artificial technologies by orders of magnitude, chemistry done safely with great artistry, and economies that fit within their ecological limits.[10] For discerning students, nature instructs about the boundaries and horizons of our possibilities. It is the ultimate standard against which to measure our use of the world.

Design, in architect Bill McDonough's words, is the first signal of human intention. The intentions behind the design of the compulsive consumer economy were to liberate the individual from community and material constraints and to thoroughly dominate nature and thereby expand the human realm to its fullest. René Descartes, Galileo, Sir Isaac Newton, Adam Smith, and their heirs, the architects of the modern world, assumed nature to be machinelike and with no limits and assumed humans to be similarly machinelike and with no limits to their wants. Consistent with those assumptions, excess has become the defining characteristic of the modern economy, evidence of design failures that cause us to use too much fossil energy and too many materials and to make more stuff than we could use well in a hundred lifetimes.

If, however, we intend to build durable and sustainable communities, and if we begin with the knowledge that the world is ecologically complex, that nature does in fact have limits, that our

health and that of the natural world are indissolubly linked, that
we need coherent communities, and that humans are capable of
transcending their self-centeredness, a different design strategy
emerges. For the design of a better society and healthier communities, in Vaclav Havel's words, "We must draw our standards
from the natural world, heedless of ridicule, and reaffirm its denied validity. We must honour with the humility of the wise the
bounds of that natural world and the mystery which lies beyond
them, admitting that there is something in the order of being
which evidently exceeds all our competence."[11]

To draw our standards from the natural world requires that we
first intend to act in ways that fit within larger patterns of harmony and health and create communities that fit within the natural limits of their regions. At a larger scale we must summon the
political will to intend the creation of a civilization that calibrates
the sum total of our actions with the larger cycles of the earth.
When we do so intend, design at all scales entails not just the
making of things; it becomes, rather, the larger artistry of making
things that fit within their ecological, social, and historical context. Design is focused on rationality in its largest sense, giving
priority to the wisdom of our intentions, not the cleverness of our
means. Just as physicians are admonished to do no harm, practi-
tioners of ecological design aim to cause no ugliness, human or
ecological, somewhere else or at some later time. When we get
the design right, there is a multiplier effect that enhances the
good order and harmony of the larger pattern. When we get it
wrong, cost, disease, and disharmony multiply.

Like any applied discipline, ecological design has rules and
standards. First, ecological design is a community process that
aims to increase local resilience by building connections among
people, between people and the ecology of their places, and between people and their history. An analog of engineering design,
which aims to create resilience through redundancy and multiple
pathways, ecological design works to counter the individualiza-
tion, atomization, and dumbing-down inherent in the consumer
economy by restoring connections at the community level. The
process of design begins with questions such as the following:

How does the proposed action fit the ecology of a place over time? Does it keep wealth within the community? Does it help people to become better neighbors and more competent persons? What are the true costs, and who pays? What does it do for, or to the prospects for, the children?

Well-designed neighborhoods and communities are places where people need one another and must therefore resolve their differences, tolerate one another's idiosyncrasies, and, on occasion, forgive one another. There is an architecture of connectedness that includes front porches facing onto streets, neighborhood parks, civic spaces, pedestrian-friendly streets, sidewalk cafés, and human-scaled buildings.[12] There is an economy of connectedness that includes locally owned businesses that make, repair, and reuse; buying cooperatives; owner-operated farms; public markets; and urban gardens—patterns of livelihood that require detailed knowledge of the ecology of specific places. There is an ecology of connectedness evident in well-used landscapes and in cultural and political barriers to loss of ecologically valuable wetlands, forests, riparian corridors, and species habitat. Competent ecological design produces results tailored to the ecology of particular localities, or, in biologist John Todd's words, "elegant solutions predicated on the uniqueness of place." Design represents the long-term effort of communities "to fit close and ever closer" into particular places and landscapes. In Jacquetta Hawkes's words, it is like a "patient love-making" between people and their places.[13] There is a historical connectedness embedded in the memories that tie us to particular places, people, and traditions—swimming holes, lovers' lanes, campgrounds, forests, farm fields, beaches, ball fields, schools, historic sites, burial grounds.

The degree to which connectedness seems distant from our present reality is a measure of how much we've lost in making consumption quick, cheap, and easy. Compulsive consumption is, in fact, proportional to the atomization of people, to social fragmentation, and to the emotional distance between people and their places. It is a measure of human incompetence requiring no skill and no wherewithal beyond ownership of a credit card.

Connectedness, on the other hand, requires the ability to converse, to empathize, to resolve conflicts, to tolerate differences, to perform the duties of a citizen, to remember, and to re-member. It requires a knowledge of the natural history of a place, practical handiness, and place-specific skills and crafts. It creates roots, traditions, and a settled identity in a place.

A second rule of ecological design is that it takes time seriously by placing limits on the velocity of materials, transportation, money, and information. The old truism that haste makes waste makes intuitively good ecological design sense. Increasing velocity often increases consumption, thereby generating more waste, disorder, and ugliness. In contrast, good design aims to use materials carefully and slowly. To preserve communities and personal sanity, it would place limits on the speed of transportation.[14] To take advantage of what economists call the multiplier effect, it would slow the rate at which money is exchanged for goods and services imported from outside and thereby exits the local economy.[15] Good design acknowledges the fact that beyond some relatively low threshold, the rapid movement of information works against the emergence of knowledge, which requires time for people to mull things over, to test results, and, when warranted, to change perceptions and behavior. The clock speed of genuine wisdom, which requires the integration of many different levels of knowledge, is slower still. Only over generations, through a process of trial and error, can knowledge eventually become design wisdom and memory about the art of living well within the resources and assets of a place. Good design aims to match the material requirements of the community with the clock speed of charity and neighborliness, which is always slower than that which is technologically feasible.

Excess consumption, on the other hand, is in large measure relative to velocity. A bicycle, for example, moving at 20 miles per hour requires only the energy of the bicyclist. An automobile moving at 55 miles per hour for one hour will burn 2 gallons of gasoline. A transatlantic flight between New York and London on a 747 moving at 550 miles per hour for six hours will burn 100 gallons of jet fuel per passenger. The difference is not just in the fuel consumed but also includes the entire support apparatus required by the increased speed of travel. A bicycle requires a relatively simple support infrastructure. An airline system, in contrast, requires a huge infrastructure including airports, roads, construction, manufacturing and repair facilities, air-traffic control systems, mines, wells, refineries, banks, and the consumer industries that sell all the paraphernalia of travel.

By taking time seriously enough to use it well, ecological design may also reset the clock speed of people's sense of propriety. The consumer society works best when people are impulsive buyers, expecting instant gratification. By moderating the velocity of material flows, money, transport, and information, ecological design may also teach larger lessons having to do with the discipline of living within one's means, the delaying of gratification, the importance of thrift, and the virtue of nonpossessiveness.

The third rule of ecological design is that it eliminates the concept of waste and transforms our relationship with the material world. The consumer economy uses and discards huge amounts of materials in landfills, air, and water. As a result, environmental policy is mostly a shell game in which waste is moved from one medium to another. Furthermore, carelessness in the making and use of materials has resulted in the global dissemination of some 70,000 synthetic chemicals, carried by wind and water to the four corners of the earth.

Ecological design requires a higher order of competence in the making, use, and eventual reuse of materials than that evident in industrial economies. Ecologically, there is no such thing as waste. All materials are "food" for other processes. Ecological design is the art of linking materials in cycles and thereby preventing problems of careless use and disposal. Nature, accordingly, is the model for the making of materials. If nature did not make something, there are good evolutionary reasons to think that humans should not. If we must, we ought to do so in small amounts that are carefully contained and biodegradable, which is to say, the way nature does chemistry. Nature makes living materials mostly from sunlight and carbon; so should we. Nature does not mix things like chlorine with mammalian biology. Nor should we. It creates novelty slowly, at a manageable scale, and so should we.

An economy that took design seriously would manage materials flows to maximize reuse, recycling, repair, and restoration. It would close waste loops by requiring manufacturers to take products back for disassembly and remanufacture. It would make distinctions between what architect Bill McDonough calls products of service and products of consumption. We do not, for example, consume carpets. We use them. But rather than send carpets to the local landfill when their useful life is over, we could return them to the manufacturer to be remade into new product. In Europe, the concept is being applied to solvents and automobiles. It ought to become routine for all products of service mislabeled "consumer durables."

The fourth rule is that ecological design at all levels has to do with system structure, not the coefficients of change. The focus of ecological design is on systems and "patterns that connect," as anthropologist Gregory Bateson once put it. When we get the structure right, "the desired result will occur more or less automatically without further human intervention."[16] In Amish communities, for example, it is the horse, which provides transportation and mechanical power for farm work, that limits what people can do. The effective radius of a horse-drawn buggy is eight miles, and its hauling capacity is low. The result is a system structured around the capacity of the horse, which serves to limit human mischief, economic costs, consumption, dependence on outside resources, and ecological damage while providing time for human sociability, a source of fertilizer, and the peace of mind that comes with unhurriedness. In the Amish culture, the horse is a solar-powered, self-replicating, multifunctional structural solution that eliminates the need for continual management and regulation of people.

In the larger culture we expect laws and regulations to perform the same function, but they seldom do. The reason has to do with the fact that we tend to fiddle with the coefficients of our problems—the rates at which things get worse or the degree to which it is permissible to poison one another—rather than deal with the structure of the systems that cause problems in the first place. The Clean Air Act of 1970, for example, aimed to reduce pollution from automobile emissions by attaching a catalytic converter to each automobile—a coefficient solution. Nearly four decades later, with more cars and more miles driven per car, even with lower pollution per vehicle, air quality is little improved and traffic is worse than ever. The true costs of that system include the health and ecological effects of air pollution and oil spills; the lives lost in traffic accidents; the degradation of communities; an estimated $300 billion per year in subsidies for cars, parking, and fuel, including the military costs of protecting our sources of imported oil; and the future costs of climate change. The result is a system that can work only expensively and destructively. In contrast, a design solution to the transportation problem would aim to change the structure of the system by reducing our dependence on the automobile through a combination of high-speed rail service, light-rail urban trains, bike trails, and smarter urban design that reduces the need for transportation in the first place.

The same logic applies to the structures by which we provision ourselves with food, energy, water, and materials and dispose of our waste. Much of our consumption, including the use of excessive packaging and the addition of preservatives to food, has been engineered into the system to meet the requirements of long-distance transport. Some of our consumption is due to built-in obsolescence designed to promote yet more consumption. Some of it, such as the purchase of dead-bolt locks and handguns, is necessary to offset the loss of community cohesion and trust caused in no small part by the culture of consumption. Some of our consumption is dictated by urban sprawl, which leads to over-dependence on automobiles. We have, in short, created vastly expensive and destructive structures to do what could be done better locally, with far less expense and consumption. Redesigning such structures means learning how politics, tax codes, and laws work and how they might be made to work with more grace and ecological resilience.

WITHOUT QUITE INTENDING to do so, we have created a global culture of consumption that will come undone, perhaps in a few decades, perhaps a bit later. We are at risk of being engulfed in a flood of barbarism magnified by the ecologists' nightmare of over-population, resource scarcity, biotic impoverishment, famine, rampant disease, pollution, and climate change. The only response that does credit to our self-proclaimed status as *Homo sapiens* is to rechart our course. That process, I believe, has already begun. But it will require far greater leadership, imagination, and wisdom to learn, and in some respects relearn, how to live in the world with ecological competence, technological elegance, and spiritual depth. We have models of communities, cultures, and civilizations that have in some measure done so and a few that continue to do so against long odds. There are still indigenous peoples who know more than we will ever know about the flora and fauna of their places and who have over time created resource management systems that effectively limit consumption.[17] There are sects, such as the Amish, that continue to resist the consumer economy but nevertheless manage to live prosperous and satisfying lives. There are ancient practices, such as feng shui, that have informed some of the best Chinese land use and architectural design for centuries, and there are new analytic tools, such as least-cost end-use analysis and geographic information systems, that will help us see our way more clearly. There are also emerging interdisciplinary fields such as "green" architecture, restoration ecology, ecological engineering, solar design, sustainable agriculture, industrial ecology, and ecological economics that may in time lay the foundations for a better world.

The problem is not one of potential but one of motivation. To live up to our potential, we must first know that it is possible for us to live well without consuming the world's loveliness along with our children's legacy. But we must be inspired to act by examples that we can see, touch, and experience. Above all else, this is a challenge to educational institutions at all levels. We will need schools, colleges, and universities motivated by the vision of a higher order of beauty than that evident in the industrial world and that in prospect. They must help expand our ecological imagination and forge in the rising generation the practical and intellectual competence that turns mere wishful thinking into hopefulness.

Stuart's letter opener came to me as a gift, an embodiment of skill, design intelligence, kindness, and thrift. Stuart used no more than one-tenth of a board foot of wood to make it. He used no tools other than a wood rasp, some sandpaper, and linseed oil. The wood itself was a product of sunlight and soil, symbolic of other and larger gifts. If I were to lose this letter opener I would grieve, for it is full of memory and meaning. Seeing it and using it, each day I am reminded of Stuart and have a refresher course in the importance of craftsmanship, charity, and true economy. I will use it for a time and someday then pass it on to another.

It All Begins with Housework

Jane Smiley

WHEN ENGLISHMEN Morris Birkbeck and Richard Flower saw Edwards County, Illinois, in 1817, they couldn't have been more certain that these American lands had been created by Providence for the specific purpose of making every Englishman a lord. The rolling, oak-studded prairie looked to them exactly like the arduously cultivated parklands of English country estates. All that was needed were some hedges, some fences, some large brick houses, and there you were—all those problems of primogeniture and entail swept away. When they made their claims, though, and began to build their settlement at Albion, they ran into a problem that Europeans and Americans complained of throughout the eighteenth and nineteenth centuries—no servants. Flower wrote in 1825, "It ought not to be concealed that we are much in want of farming laborers. . . . We are also destitute of female servants."[1] They were not alone in being so. American history, as much as it is a history of warring, settling, exploiting, and making money, is also a history of finding some way to get the work done given a severe shortage of servants. Modern American consumerism is not just a consumerism of pleasure and play; it is also a consumerism of work—housework, farm labor, transportation, industrial labor, construction. American history and American culture depart from the cultures of the Eastern Hemisphere, and from the culture of South America, exactly at the point at which Americans, who could afford to work for themselves rather than for others, were obliged to find ways to get their work done other than through the labor of subject peoples. Since the model of American consumerism now dominates the entire world, the solutions Americans came up with for the labor shortage problem have come to serve as a model of consumption that nearly everyone who has a television set desires to emulate, and they have brought us, some say, face to face with the absolute limitations of the natural world.

It's enlightening, I think, to focus upon housework. In her 1841 *Treatise on Domestic Economy*, Catharine Beecher (sister of Harriet Beecher Stowe and of Henry Ward Beecher) begins bluntly: "The number of young women whose health is crushed, ere the first few years of married life are past, would seem incredible to one who has not investigated the subject, and it would be vain to attempt to depict the sorrow, discouragement, and distress experienced in most families where the wife and mother is a perpetual invalid."[2] After this remark, Beecher goes on to detail exactly what the tasks are, in addition to childbearing, that might render the wife a perpetual invalid. They are many and daunting, and the only help (a considerable one, in fact) Beecher has to offer is a way of organizing and systematizing jobs such as soap making, dyeing, nursing, and animal care, which were then domestic responsibilities additional to the cooking and cleaning we are familiar with today. Beecher pulls no punches. At one point, she warns her reader that every woman must be prepared at all times to have her plans and arrangements interfered with and wrecked. Flexibility and a good temper are the first necessities of such a difficult life. Good ventilation and nutrition, warm clothing, and avoidance of "stimulating drinks" come right after. In other words, domestic life was an enterprise that women of the nineteenth century had to train for as athletes train today; the penalty for not doing so was illness and even death.

If we turn to analysis of the average woman's working day, it isn't hard to see why this should be so. First, there were the fires. In her book *Never Done,* Susan Strasser writes:

> Only constant fire tending—poking, shifting logs, and adding wood—could keep a hot fire going in brick ovens and fireplaces. Wood had to be felled, chopped, and carried into the house; usually men and boys cut the trees down, but much chopping and hauling was women's work as well. The job of cooking on those fires was hot and dangerous. Despite long-handled utensils, cooks had to bend and kneel and reach into the flames. Burning cinders flew from unscreened kitchen fires, skin and clothing scorched at the grate, and small children were seldom safe in their own homes.[3]

And tending the fires consumed a goodly portion of a woman's day, estimated by Strasser at three or four hours. By the late nineteenth century, fireplaces and brick ovens had been replaced by cast-iron stoves and even central heating, but fires still had to be built and tended, and the new technology was accompanied by the new danger of carbon monoxide poisoning. Strasser quotes Harriet Beecher Stowe's remark that the airtight stove "in thousands and thousands of cases . . . has saved people from all further human wants, and put an end forever to any needs short of six feet of narrow earth which are man's only inalienable property."[4]

Once the fire was built, the water had to be put on to boil, for washing and laundry as well as for cooking. A North Carolina

Farmer's Alliance organizer calculated that one female member of his audience, who had a spring sixty yards from the house, covered 148 to 220 miles per year hauling water to her house six to ten times every day, and the uphill part was with full buckets. Once she had the water in the house, she had to pour it, heat it, carry it around, and, eventually, carry it out of the house and discard it again. Although Strasser didn't calculate the portion of a woman's day spent hauling water, we may readily see why nineteenth-century women often complained that they spent their whole lives going to wells and springs and other water sources. And then, of course, there was the daily emptying and cleaning of chamber pots.

And that was before the housecleaning, laundry, cooking, and child care had even begun for the day.

In other words, although we may cherish the Federal, Queen Anne, and Victorian houses that we see in our midst and, indeed, that some of us are lucky enough to live in, we don't even begin to live the same lives within their walls that our great-great-grandmothers did. Their daily work looked much more like the daily work of village women in India than the daily work women do today.

That the servant problem was the unspoken heart of the controversies leading to the Civil War seems obvious to me. In her wise, even visionary, 1851 novel *Uncle Tom's Cabin,* Harriet Beecher Stowe, herself a wife, homemaker, and mother of seven children, is explicit about these issues. She clearly contrasts the domestic arrangements of the New Orleans St. Clare family, where Tom ends up when he is sold away from his Kentucky home, with those of the northern branch, as represented by cousin Ophelia, who comes south to visit and help. Miss Ophelia, whose lot it has been to devote herself to the raising of her brothers and sisters, "was a living impersonation of order, method, and exactness. In punctuality, she was as inevitable as a clock, and as inexorable as a railroad engine; and she held in most decided contempt and abomination anything of a contrary character."[5] Having lived for some forty-five years on one of those neat New England farms where "the lady in her snowy cap sits sewing among her

daughters as if nothing ever had been done or were to be done"—
they have already done it "in some long forgotten forepart of the
day"—Ophelia has left behind the frugality and order of New
England for the luxurious prodigality of New Orleans.[6] Although
Stowe seems fond of Ophelia, she stresses the point that
Ophelia's life has limited both her options and her character. Not
only is she a New England virgin, deprived of love and a family
life of her own, but also "her standard of right was so high, so all-
embracing, so minute, and making so few concessions to human
frailty, that, though she strove with heroic ardor to reach it, she
never actually did so, and of course was burdened with a constant
and often harassing sense of deficiency;—this gave a severe and
somewhat gloomy cast to her religious character."[7]

When Ophelia comes to live with her cousin, Augustine St.
Clare, she is explicitly contrasted with the two women most in
charge of his household: his wife, Marie, and the cook, Dinah.
Marie is a querulous invalid whose only mode of conversation is
to alternate complaints about her health and her womanly lot
with complaints about her slaves, especially Mammy, who, it becomes clear, is ruthlessly run off her feet by Marie's incessant
needs. More important is Dinah, the cook, "a self-taught genius."
On the first day of her visit, Ophelia attempts to bring order to
Dinah's kitchen, but she is frustrated by Dinah's failure to cooperate and her devotion to her own disorderly system. When
Ophelia complains to her cousin, he replies:

> Now, there's Dinah gets you a capital dinner,—soup, fowl,
> dessert, ice creams and all,—and she creates it out of chaos
> and old night down there, in that kitchen. I think it really
> sublime the way she manages. But heaven bless us, if we
> were to go down there, and view all the smoking and squat-
> ting about and hurryscurryation of the preparatory proce-
> dures, we should never eat more! . . .
>   "But Augustine, you don't know how I found things!"
>   "Don't I? Don't I know that the rolling pin is under her
> bed and the nutmeg grater in her pocket with her tobacco,—
> that there are sixty five different sugar bowls, one in every
> hole in the house,—that she washes dishes with a dinner
> napkin one day and with a fragment of petticoat the next?

But the upshot is that she gets up glorious dinners, makes superb coffee; and you must judge her as warriors and statesmen are judged, by her success."

"But the waste!—and the expense!"[8]

The real lesson that Ophelia learns in New Orleans is a lesson of the heart—that she cannot make a change in the life of Topsy, the slave child she cares for, until she opens up to her emotionally, something that her housekeeping and religion in New England have not prepared her to do. The subtext of Stowe's novel is clear in its assertion that the material conditions of domestic life in both regions create specific virtues and deficits of character, and that character dictates fate. No character is able, except by the greatest effort of courage and imagination, to transcend his or her material circumstances.

In Charles Dickens's *American Notes,* as well as in his *Martin Chuzzlewit,* the conditions of the third national region, the West, are noted as well. No one was more explicit than Dickens about the costs of western emigration to the health and well-being of the average American woman. In *American Notes,* he writes, "It was a pitiful sight to see one of these [settler's wagons] deep in the mire, the axle tree broken; the wheel lying idly by its side; the man gone miles away, to look for assistance; the woman seated among their wandering household goods, with a baby at her breast, a picture of forlorn, dejected patience."[9] When Martin Chuzzlewit arrives at the place (around Cairo, Illinois) where he had intended to settle, he finds only disease, death, and futility, especially among the women and children.

The fact is that nineteenth-century domestic life in America was not for the faint of heart. By all measures of well-being that we consider normal today, women and children in the North and West were heavily taxed and profoundly burdened by a dearth of population, vast distances, and overwhelming labor. The necessity of supplying the family's own working class in the form of children was a further onus for northern and western women; multiple childbirths resulting in an early death was the lot of many. The conditions of slaves on the plantations of the South are well known, of course, and the absolute evils of slavery infected their masters and mistresses as well, even the best of whom almost always fell short of the organizational tasks that burdened them.

The answer was contrivance. The answer was technology. After the Civil War, the answer came to be fossil fuels. All of us surely remember from grammar school the catalog of inventors and inventions that, when we were eight and ten, were of little consequence to us—Robert Fulton and his steamboat, Eli Whitney and his cotton gin, the McCormick reaper, the Singer sewing machine, the Wright brothers, Henry Ford. Was anything invented anywhere else in the world? Yes, indeed, but these grand American inventions were the children of the same necessity that gives us the antique, fascinating, and sometimes mysterious domestic gadgets we find in attics and museums—cherry pitters, apple peelers, corn shuckers, strange rocking benches where the seat is fronted by a railing (for rocking the baby and keeping one's hands free at the same time), eggbeaters. All these gadgets and thousands of others were designed to be savers of time, or enablers of a woman's doing two things at once. It was obvious. If there was to be no reliable class of workers who could be made to carry buckets of water day in, day out for their entire lives, then some mechanical system had to be devised for carrying the water and doing everything else. There was no such class of workers, and technological progress in the United States did not meet with the Luddite resistance that it met with in Europe. Those who were being replaced by machines in the United States had higher ambitions anyway.

If we remember nineteenth- and early-twentieth-century feminism, we usually think of it as having had to do with women voting, but feminist theorists had a lot to say about domestic drudgery as well. Their desire to empower women was task specific and particular, and their feminism grew out of a thorough knowledge of the way most women spent their time. Catharine Beecher's works exemplified the strand of feminism that hoped to maintain the separation of male and female worlds into the public and the domestic but aspired also to raise the status of women's work and their self-image. Other theorists saw that many domestic tasks of the eighteenth century, such as candle making, spinning, and weaving, had moved out of the home. They had high hopes that this pattern would continue and would soon embrace the most onerous task of every woman's week: doing the laundry. What with water hauling, fire making, natural-fiber fabrics, voluminous clothing, and serious dirt, the weekly laundry was a heavy, dreaded, all-day labor. Clothes had to be soaked, blued, soaped, bleached, boiled, rinsed, wrung out, hung out, starched, and ironed, and several of these operations had to be done more than once. Even Beecher, who believed that system and good temper could lighten most domestic trials, advocated help with the laundry, in the form of either washerwomen who would come in on laundry day or, preferably, washerwomen who made a living from doing laundry every day of the week and would receive at least part of the family laundry—sheets, tablecloths, men's clothing—each week. This passing of laundry out of the home, called by Strasser the movement of a domestic task into the "craft stage," was followed, as were ginning (Whitney cotton gin) and spinning (spinning jenny) and weaving, by an industrial stage, which was well under way, in both urban and rural areas, by the early twentieth century. The washing machine had by this time been invented in various models, so centralized laundry was a natural—working women went out to wash what their sisters went out to create. But whereas the making of clothing moved, and has continued to move, farther and farther away from the individual American home, laundry, thanks to the domestic washing machine, moved right back in during the middle third of the twentieth century (just as centralized transportation in the form of railroads and streetcars was reindividualized in the form of automobiles at about the same time and for many of the same reasons).

Theorists of domestic history argue about the relative losses and gains brought about by individualized domestic laundry (the woman, for example, who does her own laundry at home loses a traditional opportunity for social contact), but no one in her right mind cares to return to the torments of nineteenth-century clothes washing.

The lesson to be learned here is that feminism and consumerism are tightly linked. The freedom to be educated in something other than home economics and the freedom to earn money, to have a vocation, to have an avocation, to engage in all the useful and useless activities of our historical moment, depend on the lightening of the domestic load through contrivance, technology, and the use of nonhuman, nonanimal power. The key observation of our time, it seems, is that others all around the world wish to have this freedom as well, and the key question is, How can they get it?

These are, at heart, political and moral questions, and they demand political and moral answers. I have gone through this brief history because I don't believe that the environmental movement addresses these questions very often or in a manner that is informed by knowledge of how we got here. The natural world sets a limit (though, so far, we don't know exactly what that limit is) on our use of it to do our work. Most arguments seem to be about where that limit should be and damn the consequences. In my view, the arguments should be bolder and more honest. First, we should discuss whether there *should* be winners and losers in the bid for natural resources, then we should discuss how those categories would be decided; and then we should discuss the limits for the limited and the prizes for the lucky. Finally, we should admit that even though these issues are almost never decided through discussion or decided ahead of time, given the selfishness and shortsightedness of most human groups, such discussions might reveal us to ourselves and help us, as individuals, sort through who we are and what we want.

Those who have had the privilege of free time or extra goods in the capitalist, colonial, and imperialist world of the past 350 years have generally not doubted their right to it. What was not "God given" was at least "according to the natural order"; what was not inherited and sanctioned by tradition was earned; what was not blameless good luck was harmless incentive to the incentiveless. At any rate, the generosity of the natural world afforded considerable leeway; for much of this period, the globe has seemed to be an open, spacious place, easily accommodating the human urge to move off and find new territory. In several ways, the colonialists had considerable luck. It is estimated that as much as 90 percent of the New World population melted away before the con-querors from Europe ever saw them, victims of epidemic diseases to which, in their isolation, they had never been exposed. Another bit of luck was European topography, which, unlike either Asian or South American topography, encouraged the pasturing of ani-mals for food, work, travel, and war, giving Europeans all sorts of experiences that they were able to put to good use in conquering other populations. Christianity, with its history of active conver-sion, its absolute conviction of righteousness, and its placing of human concerns at the top of the earthly hierarchy, was another piece of luck. The desanctification of the world (a necessary pre-condition for capitalist accumulation, according to economist Robert Heilbroner, and for the development of science) was al-ready a fixed tenet of Western thought before the Middle Ages. We may look back and see that when the Europeans set out for the rest of the world in the seventeenth century, there wasn't much to stop them. They had all the convictions, the manpower, the practice, the inventiveness, and the motivational greed they needed, and problems of technique and distance were but chal-lenges and goads.

North America seemed, in the nineteenth century, both the natural flowering and the revolutionary opposite of European life. On the one hand, here, many Americans thought, true Christian ideals were being lived out and the natural superiority of the Caucasian race was being demonstrated. On the other hand, the dead and weighty hand of the European class system had been thrown off, and every man could find the space to realize his true worth. The inevitable outcome of these contradictory reactions to European traditions was the American philosophy of the self—an ever expanding sense of individuality, no man to be treaded upon by the tyrant, expanding into a resistance to all relationships be-tween men that were the least hierarchical. The loner, the cow-boy, the outlaw, the do-it-yourselfer, the farmer or rancher: whether or not they fail or succeed, at least they can declare, "I did it my way." An alternative model—since some things, such as railroads and the Brooklyn Bridge, can be produced only by groups—is the land baron, the robber baron, the industrialist, the ruthless entrepreneur, who declares, "They did it my way." No one aspires, however, to declare, "I did it his way."

The sort of individualism tempered by social responsibility that was the vision of the framers of the Constitution and the authors of the Federalist Papers came under considerable pressure with the expansion of both American territory and American population. Some groups were able to maintain social cohesion for some time, but most groups not sustained by strict religious doctrine succumbed to the temptation to fragment, a temptation strengthened by accepted social ideals such as exploration, the quest for success and novelty, and the value given to desire rather than obligation. Moral values such as "being true to oneself" and "letting your conscience be your guide" reinforce the requirement that the individual look to himself first and last, even if, in the end, he is isolated in his integrity.

These sorts of ideals, in both their good manifestations and their bad ones, grow out of a material culture in which an individual is aided in his or her work and play by some other source of power. The simplest image is the cowboy on his horse—his wants are simple, but he does need the horse; otherwise, he can't be defined as a cowboy. The do-it-yourselfer needs a workshop full of tools, some manuals, and a good supply of batteries and electricity. The loner never goes away on the bus—the moment he gets off the bus is the moment that defines him as a loner; the moment he gets into a car of his own is the moment when we know whether he will live or die. The outlaw, likewise, has no getaway subway car—his very identity depends on his having a car of his own or stealing one. The farmer is out there alone on his tractor; the rancher, in his pickup truck. All are kin to the housewife, alone at home with her washing machine, choosing to wash once a week or every day, mixing whites and colors as she pleases, using Lux Flakes, Wisk liquid, or new, blue Cheer. The robber baron, too, is defined by the power he harnesses—the large machinery spewing noise and exhaust, the vastness of the project, the antlike swarms of men scurrying about its base, the distances the raw materials have traveled, the unprecedented expense of

everything. (His own conspicuous consumption, a defining feature of European privilege, is beside the point. We may remember something Louis XIV wore and where he lived, but we don't
remember what John D. Rockefeller wore or what his most expensive residence looked like.)

Perhaps the ideal that all Americans share by reflex is that of
freedom. The actual definition of freedom runs the gamut, but
most people conceive of it as the experience of choice or, at least,
potential choice. It is also free time, freedom to spend some discretionary income, freedom to state one's opinion, freedom to get
away rather than the obligation to stick around. I think that most
women would agree that in these areas, we of the 1990s are freer
than women were a hundred years ago. We may choose whether
or not to have children and how many to have. We may choose
whether to raise them entirely ourselves, hire a nanny, or take
them to day care. Without fires to build, water to carry, chamber
pots to empty, and food animals to care for, we may choose how
clean our house is, how well fed our families are, how much time
we spend at these activities, and how much time we spend at
home and away from home. More important, the ebb of housework has allowed women to choose other work, as doctors,
lawyers, novelists, professors, firefighters, factory workers
(though for much of the nineteenth century, this was an option
for only some women, in some parts of the country), bus drivers,
and all the other jobs and professions that we see around us.
There is no longer a social need to focus all the energies of one
class of workers on a particular type of work. Therefore, the sort
of attention to what women "should" be doing, "were created" to
do, and are "morally obligated" to do is not as much a feature of
our culture as it was in the nineteenth century, when the number
of ways in which women were defined as domestic was surpassed
only by the number of times they were urged to be so.

I live in a clean and pleasant house with sizable rooms and
plenty of light. It is heated by natural gas. When I get up at a
quarter of seven, I roll over and turn on the light, the work of
a moment. It takes me fifteen minutes to make my children
a breakfast of eggs, toast, fruit, and cereal and another ten to

make my son's lunch for school (peanut butter sandwich, string cheese wrapped in plastic, fruit leather, a drink, chips, and graham crackers). While the children are eating, I put in a load of laundry and walk around the house, making beds and picking up clutter. I unload the dishwasher, load up the dirty dishes, take a minute or two to feed the dogs their kibble, help my son get dressed, and say good-bye to my daughter when she leaves for school with her car pool. At a quarter past eight, my son gets on the school bus. I also perform some entirely unmechanized work—horse care. I scoop up feed, throw out hay, clean stalls. By half past eight, my house is relatively neat, my animals are taken care of, and my children are away for the day. It is time to be a novelist, to pursue my avocation of horseback riding, and to communicate with friends by telephone.

My friends are an administrator of a nonprofit music festival, a lawyer, a college professor, an ocean geologist, two horse trainers, a horse masseuse, a people masseuse, a woman who has no job outside the home, a waitress, a racetrack exercise rider, a publishing executive, and a student. All are women. None of us finds her housework load onerous. All of us rely entirely on appliances, electricity, supermarkets, automobiles, schools, systems for the delivery of natural gas, telephones, and television (for at least occasional entertainment of our children). Most of us are in our late forties to mid-fifties, and all of us are in excellent health. I have borne the most children, three, and all of our children are being, or have been, raised by the women who gave birth to them. All of us have full and busy lives, with plenty of stimulation and plenty of choice. All of us have some leisure time. None of us forgoes sleep or nourishment owing to lack of time, energy, or money. Our lives, in general, if not in detail, have turned out much as we'd hoped they would—our work is satisfying, our home lives are relatively peaceful, and our homes are reasonably clean and are well heated and well lighted. The question is not how we could make better use of our time, since no one makes perfect use of her time (and the American ideal of freedom means that we are free to waste our time if we wish to). The question, rather, is what it would take to make us give up the privileges and freedoms we have, and return to a shorter and more laborious, repetitive, un-healthful, limited, and worrisome existence—or, since most of us have daughters, what it would take to persuade us to shorten and limit *their* lives.

I have conducted an informal survey on this question. My daughter would accept $2 million per year to limit her career choice to housework and to give birth to seven children by the time she is thirty. Under no circumstances, however, could she be bribed to give up plumbing, electricity, automobiles, running water, or supermarkets. The publishing executive would consent to live as her great-grandmother did in exchange for being guaranteed 200 years of life so that she could subsequently live as she does now. She would also agree to live as her great-grandmother did in order to avoid instant death. My friend the people masseuse, however, declared that instant death, as an unknown and possi-bly interesting experience, would be preferable to living as our great-grandmothers did. One of the horse trainers, a woman ac-customed to repetitious routine and hard physical labor, thought that she could accept her great-grandmother's lifestyle (including the bearing and raising of twelve children) if the alternative were physical or mental impairment. The professor could accept it if the alternative were the deaths of her children. My cousin, who as an actress and office worker, lives a more difficult life than do some of the rest of us, thought that $2 million plus life in a beau-tiful place might be compensation enough, but she was suffi-ciently drawn to such a life to prefer it to terminal illness, physi-cal impairment, or instant death. Only my friend the music-festival administrator, who is the oldest of us and lived for twenty years in the English countryside, was willing to seriously entertain (for herself but not her daughters) the idea of returning to a nine-teenth-century female life, and then only if it could be lived in a village, with plenty of company and mutual aid. We agreed that such villages no longer exist in America, if they ever did. And those who recall small-town life in its heyday (for several years around the beginning of the twentieth century) always recall its provinciality, narrowness, small-mindedness, and gossipiness as much as they recall its coziness.

The fact is that the way we live our lives today reflects what our ancestors aspired to get away from. Housework is easy for us because it was once so thankless and unremitting that the methods and products designed to ease it found an eager market. People use cars and space to get away from one another because they have found isolation preferable to conflict. Americans choose desire over obligation, television over conversation, arduous medical care over death and disability, and high earnings over strict frugality because they *may* do so, not only because they are encouraged to do so. Household drudgery, staying in one place most of the time, close social conditions, homemade entertainment, disability, early death, and habits of thrift smack of deprivation as much to those who haven't experienced them as to those who have. Those who advocate them, such as Wendell Berry, are seen as not only unrealistic but also curmudgeonly. There is much talk of the emptiness of modern life, but think of emptying chamber pots of the accumulated waste products of seven or eight household members every day for the rest of your life. Think of asking someone else to do it. Think of shirking your duty and not doing it.

Nevertheless, the world's 5 billion or more people cannot live as Americans aspire to live. The inventive solutions to the challenges of nineteenth-century American life powered the American phase of the history of capitalism, just as previous innovations powered earlier phases in other locations—the rise of banking in Florence, overseas exploitation in Amsterdam, colonial settlement in England. In each case, a new center, with a new solution to its own problems, superseded the innovations of the old center and transformed the world; then it was in turn superseded when the solutions it had found were undermined by their own internal contradictions. Just as the British Empire grew too large and unwieldy to hold together, so American technology has grown to be a nearly unbearable ecological burden. The next phase of capitalism must and will solve this problem. It will take place somewhere else, not in the United States, and its social and economic organization will be something that we in the United States may not recognize as positive and certainly will not feel comfortable with. But eventually, just as the British in the 1980s reorganized themselves along rather American lines, we will reorganize ourselves in imitation of the new successful society, the new center of capitalist enterprise.

My cousin the actress imagined how she might bring herself to live as our great-grandmother lived when she emigrated to the United States from Norway. She said she might be able to "Zen" it, that is, focus on the spiritual aspects of her given task. Her solution was not the same as the one Miss Beecher advocated for nineteenth-century women—cultivate patience and flexibility and look to your future reward—but was, rather, a discipline of being present in any given moment and taking pleasure or satisfaction in the components of the task. Miss Beecher's notions of daily female self-sacrifice for other family members have come to seem suspect, but perhaps we can cultivate a daily concentration, a moment-by-moment fullness in our awareness of ourselves and our world that makes drudgery something else. This notion was seconded by the horse trainer, who practices, with every stall cleaning, every measured step and gesture, a personal discipline that accommodates the horse's relatively intransigent and inflexible nature as a large, easily frightened, and opinionated animal.

On the one hand, we Americans have become who we are for specific historical reasons, and our solutions to the challenges presented by a small population and a large landscape still have a certain appropriateness, mostly because they are still marginally affordable through ever accelerating inventiveness. We live widely dispersed, connected by highways and cars and telephones and computers and airplanes. We have built these large systems and are, for the most part, stuck with them. Lives lived in resistance to them are much more difficult and eccentric than those lives would be if we didn't have such systems. More important, we Americans still feel our landscape to be large and sometimes overwhelming. Thunderstorms, tornadoes, hurricanes, blizzards, earthquakes, fires, power outages, oceans, mountains, floods, droughts, hailstorms, sudden freezes. Bobcats and mountain lions, Lyme disease, avian flu, whales grounding, hundreds of seals herding millions of fish to their deaths and filling the air with the stench for months. We haven't lived on this continent long enough to decide whether we are dominating it or it is dominating us, and we live in active contention with the environment in a way that others in many parts of the world do not seem to. They have lived long enough with their monsoons, their earthquakes, their permafrost, or whatever to submit, accommodate, accept. Our attack on nature still works, and it has many adherents. It has done exactly what it was intended to do—promote individual existence above all other things. Even as we deplore this, we enjoy it; the healthy, stimulated, communicative, free, active life of desire feels good enough, most of the time, that we fear to give it up for something unknown and, perhaps, unknowable.

On the other hand, those Chinese, Thais, Japanese, Indians, and others who bring their different histories and much larger populations to our model of thinking and consuming must transform our way of doing things into something else. Just as Americans stole English inventions in the early nineteenth century and then outdid them, these populous and inventive societies must learn from our mistakes, must see that our solutions can't work for them, because nature will set a tighter limit on them than it has on us. But the fact that they must eventually do so doesn't mean that the path from here to there will be straight, smooth, rational, or humane. They will make mistakes and choices that they will shrink from and then deplore, akin to our choices of black slavery and Native American genocide, which we now recall with horror—choices that Europeans even then were disgusted by, as we will be disgusted by what is to come.

Every culture produces a solution to its challenges that draws on and affects the whole culture. Just as the material opportunities and drawbacks of the American landscape produced both technological innovations and cultural ideals, so, too, will the future center or centers of capitalism produce solutions that simultaneously resolve physical difficulties and promote cultural attitudes. We may suppose that technologically based individualism will give way to something else and that in Asia, these ideals will draw on a significantly stronger communitarian tradition than Americans have. We may suppose that Asian societies will ritualize relationships, life passages, and daily life in a way that Americans do not care to. We may suppose that attitudes toward the natural world that are evident in Asian art of all kinds will operate as a brake on the wholesale commodification of landscapes and natural resources. We may suppose that individuals will bring a kind of Zen-like discipline to their daily work and pleasure. We may suppose that what we see as terribly missing in our own culture as it develops and exaggerates itself in its late stages will be answered by the next culture, just as the class society of English culture that Birkbeck and Flower sought to import wholesale to Illinois was answered by the nascent individualism of American culture. We may suppose that the peculiar charms of our own culture, the way our culture twists the kaleidoscope of human nature and presents a temporary pattern of reality, will be lost, and missed. How these general expectations will work out in detail we cannot imagine, however, because it is the nature of capitalism to present an utterly new solution in a new location when old solutions have collapsed upon themselves.

The history of housework tells us that consumerism was once a solution to the problem of labor shortage. The problem of labor shortage in America had been, in turn, a manifestation of the solution, in Europe, to the problem of a fixed class system. Our current problem of ecological deterioration will find a solution, too. We may like it, and then again, we may not.

Equipoise

Martin E. Marty

UNIVERSITY OF CHICAGO economist and political scientist
Marvin Zonis has posed well the good and bad news of the global
culture at the turn of the millennium. The good news is that the
market has won. There is now a global market, marked by a business economy. It offers many human goods and many goods for
humans. (My gloss on this: after the spinning out of millions of
miles of barbed wire to surround camps, after the shedding of
rivers of blood and the spilling of oceans of ink to force ideologies
of communism on populations, the antimarket systems have im-
ploded. They leave behind only a few regimes that call themselves

communist, and these are devoted more to holding power than to expounding belief. In places such as China, the more or less free market, even if without accompanying political freedom, is coming to prevail.)

The bad news, says Professor Zonis, is that we—meaning, no doubt, the major actors in the global economy, with the United States in a key role—have not the faintest grasp of a social philosophy to animate, monitor, and inspire this market. (My comment on this—do not hold Zonis responsible for it—capitalism, as the polar alternative to communism, is the name often given to the cluster of ideologies and systems that have outlasted their pitiless rivals. Certainly, capitalism, however described, brings with it social implications; and, more than many economists would allow, in a way each particular expression of it *is* a philosophy. Whether the system is sufficiently definable and coherent to be of much help, whether it is a sufficient social philosophy, whether it is a sufficient address to the personal search for meaning and morals of the individuals who make up societies, remains open to question. But it has won.)

All humans must consume, whether what we use or use up is a renewable resource or not. We eat and drink. We chop down trees or mold clay into bricks to produce shelter. We grow cotton or shear sheep or, more frequently in recent times, invent and exploit plastic materials to provide cloth for our clothing. To raise the question of whether we are consuming our future or to make the observation and then the judgment that we are doing so, is to apply a negative coating to the concept. To add the *-ism* is to reduce it to almost pure negativity.

I will make every effort to do some justice to these nuances in the discussion that follows, although such an effort cannot be satisfying when one tries to traverse the no-man's-land between polarized parties—in this case, those who see few or no problems in consumption and those who see many problems, who indeed foresee nothing but destruction, in most current patterns of consumption.

To use words such as *monitor* and *judgment,* as I have already done, is to imply that the world of consuming has moral implications. Even to employ them on minimal terms will rouse some economists to take a defensive stance: economics, for them, is a matter of mechanism and not morals, market laws and not ethics. Whether or not the market can be thus wrenched from the human context wherein motivations in respect to allocation of resources are of acute concern, demands moral inquiry. If and when the consumption is harmful to the existing environment, as in the case of what excessive emissions of carbon dioxide do to the air we breathe and the temperatures we endure, it is hard to escape the notion that morals are involved. And when the contending parties argue whether consumption patterns reflect chasms between rich and poor, as, of course, they do, those who are responsive to issues of justice will see consuming as an ethical frontier.

After pointing to such realities, one must respond to the questions about what will come into play. Whose morals? Whose ethics? What approaches and systems shall prevail, or is there to be only a wildly free market of options, with endless and possibly fruitless argument over them? The United States and Canada are always described as pluralist societies. Translated to sports terms, this means that any number can play; many do; there are some explicit rules of the game; and these are surrounded, supported, and tested by the various habits and customs that come with the ever changing culture. By *number* is meant the variety of groups devoted to sometimes complementary but often competing religions, philosophies, interests, and ethics. What it is to which people commit themselves and off which they live will inform what they think about their consuming. Conversely, those who uphold materialist philosophies will say that what and how they consume will determine that to which they commit. Thus from Ludwig Feuerbach, an antecedent of Karl Marx, *"Man ist was er iszt."* "One is what he eats." Be wealthy, afford luxury, and you will very likely be a proponent of gross consumption. Be poor, be deprived of lavish goods, and you will tend to call for restriction of consumption in rich countries or among rich sectors of society. Unwilling even to try to settle the probably unsettleable debates over the materialist issue in ideology, we can act on the observation that North American citizens, even when some of them argue against connecting morals with economics and hence with consuming, do moralize, especially when they engage in mutual condemnation. Whoever has an ear open to American pluralists will not lack evidence of moral concerns on all sides.

Still, the question haunts: Whose morals? Whose ethics? Most cultures and societies develop some coherent and observable answers to such questions. Mention the Cosa Nostra and you will conjure up images of godfathers and Mafia ethics. Augustine of old enjoyed observing the patterns of moral authority that even the worst bands of thieves developed—and used them to see analogies in politics. There are Beltway interests in the culture of the nation's capital. Champions of consumption know exactly what they mean when they speak of the green people's ethics. The "greens" return the favor with accounts of wasteful extravagance and grossly conspicuous consumption among the free-enterprising and free-spending rich. In almost all debates, it is assumed that there is a "culture of poverty" that animates our lives with certain moral patterns and that on a global scale, North and South live in accordance with different ethical motifs. We generalize about the consumption of forests in less-developed nations, where farmers burn trees to create fields so that they can plant and harvest food to consume, or in cold and war-torn places such as Bosnia, where people burn firewood to survive.

Are there American repositories or resources on which our complex cultures and societies draw to face the moral issue? Some are discernible to those who study our histories. For reasons of efficiency and familiarity, I deal here with the United States, confident that many of the features I describe also survive in Canada, given the two countries' shared European heritage and their similarities in the when and where of settlement and development.

These traditions are made up of ideas and practices alike. Bringing them up underscores my belief that even in diffuse and ordinary behavior patterns, ideas do have consequences. Philosopher Alasdair MacIntyre needs only one illustration to alert those who might slight such a notion. He tells how Thomas

Carlyle was confronted by a man of business who snorted, "Ideas, Mr. Carlyle, ideas, nothing but ideas." In MacIntyre's telling, Carlyle answered, "There was once a man called Rousseau who wrote a book containing nothing but ideas. The second edition was bound in the skin of those who laughed at the first."[1] Such an instance proves nothing but signals much. It is our purpose to see ideas in an environmental, in this case, national, context.

That environment, of course, compromises the formal sets of ideas, or places, with some that are more related to informal practice. One exhausted academic who devotes himself to this theme writes:

> What is most impressive about market values from a religious perspective is not their "naturalness" but how extraordinarily effective and persuasive their conversion techniques are. As a philosophy teacher, I know that whatever I can do with my students a few hours during a week is practically useless against the proselytizing influences that assail them outside class—the attractive (often hypnotic) advertising messages on television and radio and in magazines and buses, etc., which constantly urge them to "buy *me* if you want to be happy."[2]

One must grossly oversimplify when dealing with the scale of the issues in a brief space. Here is an image: the late social philosopher Ernest Gellner liked to say that complicated societies live in accordance with formal and informal "approved social contracts." These draw indefinitely on events and arguments that have occurred at especially creative or crucial times. The heirs of the originating and decisive events and actors may rework the ideas and ideologies almost to the point of making them unrecognizable. What sixteenth-century Calvinist or seventeenth-century Puritan would recognize what has evolved as a transformation of the Protestant ethic? Still, unless there be total revolution in a civilization, contemporary citizens will reflexively and reflectively live off deposits and experiences from the past. Gellner's image is of a glacier slowly and almost imperceptibly moving down a long and rugged mountain slope. It will pile up debris and ice behind a ledge and then move over what he calls a "hump of transition" to a new, approved social contract. But this momentous move leaves behind what Gellner compares to an ideological and practical moraine. That figurative glacial moraine determines a new landscape. One can move some boulders, do some new kinds of planting, and in other ways adapt, but the moraine remains.

Three moves by American people over the hump of transition have done more than anything else to inform the economic ethics and form the moral inquiries and arguments about consuming.

First is the colonial and subsequent biblical heritage, which refers to the recourse American majorities choose in letting the Hebrew scriptures and the New Testament, variously interpreted, have a bearing on their concepts of the morality of consumption. They do not lack texts about community, the meaning of work, the use of material goods, and the obligations one has to others. The Prophets and the Gospels are rich in reference to these.

Concurrent with the formation of the United States, a second complex of ideals held sway. Call it the Enlightenment "with a big *E*," as historian Crane Brinton refers to this philosophical and religious movement. With it came new philosophies of the individual, of freedom, of allocation of resources, and of markets. These outlooks are often openly in conflict with biblical injunctions and promises, even if both sets are held by the same people without their showing evidence of split personality or cultural schism.

If the biblical and Enlightenment agents and events left their moraine deposits and did their shaping work late in the eighteenth century, in the late nineteenth century a third impetus became apparent as North Americans responded to the challenges and opportunities of industrial, corporate, and advanced technological situations.

If one were assigned the task of sculpting on a Mount Rushmore visages celebrating consumer economic thought, three nominees could be Jesus (with the prophets and apostles carved into a garland as background); Adam Smith (with John Locke, Benjamin Franklin, and others of the European and American Enlightened ones in a laurel frieze behind his head); and Charles Darwin, with some barons of industry behind him.

That third instance is too allusive to be condensed into a parenthetical phrase. Here, it is coded into what came to be called social Darwinism, which is not all that compatible with pure Darwinism, whatever that is. It refers to the elevation of the competitive principle to absolute status in the governance of economic decisions and outcomes in a society. Behind Darwin's bearded head, one might then want to carve figures such as Andrew Carnegie, the magnate who published *The Gospel of Wealth*,[3] a quasi-religious interpretation, or John D. Rockefeller, who wrote of the American Beauty rose. Such a flower, he argued, acquired its beauty and survived because lesser blossoms along the stem were pruned so that as the "fittest," it got what it deserved: the main, or even all, of the nurturance.

Bring up any issue having to do with the morals of consuming—and the possible immorality of overconsuming—or of being unconcerned with the fair distribution of human goods or the lack of concern about the future as natural resources are depleted, and you will hear Americans both casually and formally drawing on resources of their spiritual "moraine." Whoever spends time assessing what those resources entail and what they demand will soon realize that these are often incompatible.

The Hebrew prophets may have been what one wag described as the party out of power criticizing those in power, but they were also more than that. They shared the covenantal claims of Israel yet saw the leadership and practices of their contemporaries being contradictory to those claims. Especially, inequality among the people offended them, and showy consumption outraged them.

Isaiah, who usually attacked men, was also gifted at satirizing the showy women of Zion who had "tinkling ornaments about their feet," plus "the chains, the bracelets, the bonnets, and the ornaments" that went with this style (Isaiah 3:18–23). Amos criticized them as "the kine of Bashan," "which oppress the poor, which crush the needy; which say to their husbands, 'Bring and let us drink'" (Amos 4:1).

In their tradition but now at the head of the Christian testament are the words of Jesus as presented in the Gospels. Jesus

and his disciples may have chosen the simple ways, and he described himself as homeless even as he sent out his followers with
minimal resources and demanded that they seek and use no more
than that. But his parables and sayings suggest that his and the
Gospels' audiences tended to be middle class. They knew what
proper garb and etiquette at banquets should be. Yet Jesus' words
show him always upsetting social and economic arrangements.
He posed most direly the spiritual and the material devotions: you
cannot serve God *and* Mammon, the personification of money, of
the material.

Against these pose Adam Smith, himself a moral philosopher,
who wrote much on morals and political economy, explicitly in
The Theory of Moral Sentiments. He took over from Christianity
its dark appraisal of human self-interest but then spun it into
something positive:

> It is not from the benevolence of the butcher, the brewer, or
> the baker, that we expect our dinner, but from their regard
> to their own interest. We address ourselves, not to their hu-
> manity but their self-love, and never talk to them of our own
> necessities but of their advantages.[4]

The very nature of people as self-concerned, Smith asserted,
could become the engine of their productivity and the motor for
their contribution to an increase in invention and achievement.
More goods would be produced—even if inequitably distributed.
How does this square with Jesus' view of self-interest and self-
centeredness; his concern to be free from the allure of the very
possessions and goods that Smith, Locke, and their Enlightened
compeers cherished; his interest in seeing the rich—with whom
he sometimes consorted—turned away while the poor came to
possess the Kingdom?

Prophetic Israel had a communal ethic for the Q'lal Yahweh,
the congregation of Yahweh. The Jews who became the early
Christians thought in terms of *koinonia* and *ekklesia,* the intimate
and integral fellowship of those "called out." (Let it be noted,
however, that their experiments with a communal economy described in the Book of Acts were imperfect from day one, seen to

be ineffective through figurative day two, and abandoned by day three.) But to render the modern competitive principle—the notion of survival of the fittest, and free choice by the haves to consume resources at the expense of the have-nots or have-lesses—as being compatible with the early believers' understanding of community demands acts of imagination and mental wrenching that leave one weary.

To do so is to magnify an element in the practical American translation of Adam Smith–ism, as observed in the 1830s by French visitor and wise commentator Alexis de Tocqueville. He saw the value of self-interested action among the citizens: "It disciplines large numbers of people in habits of regularity, temperance, moderation, foresight, and self-command." But it led to an isolating individualism, to the citizen who "leaves society at large to itself." (This may anticipate something of what Marvin Zonis saw had happened in a setting that lacked a social philosophy to go with the victory of the market.) Tocqueville, again: each individual in self-interest is led "to sever himself from the mass of his fellows and to draw apart from his family and friends," to say nothing of drawing apart from society as such. "Individualism at first only saps the virtues of public life; but in the long run it attacks and destroys all others and is at length absorbed in downright selfishness."[5]

So how do Americans live on a landscape with such moraine deposits as features and resources, jarringly different as they are from each other? That they do so is apparent. That societies are capable of living with relative incompatibilities is obvious. I think, for instance, of Will Herberg, who at midcentury began to detail a "civic," or what was later called "civil," religion. He argued neoconservatively that any functioning republic needed such a religion and that America was fortunate to have such a creative version. Then he went on to stand in the Hebrew prophetic tradition, which he preached and admired and to which he would be loyal, to say that from that tradition's angle, America's necessary and salutary civic faith or religion was idolatrous. Idolatry was the worst sin in the prophetic vision. Herberg saw America committing it and, when talking about civil religion, in effect said never mind, or do not let minding deter you from being loyal to both altars.

Societies do what individuals do: live with what have been called variously "universes of discourse," "provinces of meaning," or "arrests in experience" that are called "modes." Ask an American what is his or her most profound concern in life, what is "the whole ball of wax," by what does she live and what would she want her children to live by; what is her biggest daily and long-term practical interest? Off guard, most will say, the economy. "Free enterprise" is the concern. "The American way of life" along with it is the whole ball. "No new taxes." "The guarantees of Social Security." "Enough money to send the kids to college and to ensure a prosperous time toward the end of life."

One second later, ask, "Do you believe in God?" On guard, most—more than 90 percent—will say yes, God concerns them ultimately. They would live by the stories of God, and they want their children and society to live by the commandments of God. They want the young to be brought up morally in some sort of faith and want to be right with God at the end of life. Remind the interviewee, "But a moment ago, you were saying all those things about the economy, not about God." The new reply: "But a moment ago, you were not asking me about God." The interviewer moves on, bemused, while the interviewee blithely continues living the rest of life without too much sense of dissonance and contradiction, where there should be some if the modes in our mind, the universes of our discourse, were all logically interrelating.

Not for a moment need the religious suggest that only their fellow believers are or can be moral in the field of economics and consuming. Very often, those most responsive to prophetic traditions, who advocate styles of living that they argue come more nearly into congruence with prophetic calls, spend their time judging their fellow religionists for being institutionally self-interested. For selling out to the culture when they should be a counterculture. For aping the world's ways with their devotion to market analysis as they determine the location and styles of their places of worship, the music they use and the products they sell, the justification of their churches' and their own devotion to "consumer capitalism" or by whatever other epithet it gets called. On such occasions, many of the religious will find true companionship, if not communion, with the other-believer and the nonbeliever.

Meanwhile, many personally religious devotees of a consumption limited by nothing but the free choice of those in a position to consume criticize the "green" and "ecofeminist" and "ascetic" moralists, who they perceive as condescending. They look for and advertise the manifest contradictions between the pronouncements of those who would restrain consumption and the consuming activities of the same. They scan the ancient texts to find the more rare justifications for parts of their chosen way and outlook. (W. C. Fields: "I have spent a lot of time searching the Bible for loopholes.") These free-enterprisers deride advocates of the social gospel or of other communal ethics that would use religious messages and government support to limit free enterprise and free consumption.

So two sets of claims are in violent conflict, yet life has to be lived. Faiths and philosophies are in opposition, yet consuming must go on and overconsuming or unequal consuming will occur. What does someone who would stand in the prophetic and perhaps the Gospel traditions do to effect change in thought and action?

The first thing that occurs to those who march or stumble out to the spotlight when issues of consumption arise is that their way of living will be subjected to scrutiny. When invited to contribute to this volume, I immediately reminded the inviting editor, who knew me well enough, that I was a complicitor in consumption. My Victorian house is larger than it need be. Carbon monoxide is belched from the exhaust pipes of my family's two cars. Although our tastes are relatively simple and we have no objects that the luxuried would call luxurious, we consume disproportionately to the way the rest of the world does. I am upper middle class, a participant in two pension plans, and have Social Security and more kinds of insurance than anyone needs. How can I be a credible commentator on consuming? Where is the place to stand to view the world?

Even those who bring more rigor to their ways had better know that under the spotlight, their warts will show. If advocates of the simple life get their message out by using a word processor; if they have been jetting by whatever class to conventions in conventional convention hotels with all their luxuries; if they have produced books that get advertised and marketed; if, having gained royalties, they have enjoyed anything more than a roof, a piece of bread, and a cup of water, it will be a compromised message—if compared with the consumption possibilities of most of the world.

Add to all this the complicating word that to most religions and in many philosophies, asceticism and renunciation of worldly comforts and goods are practices for the exceptional few who choose to follow the call to utter simplicity. The same religions confront the ascetic with counterclaims about what they are denying, the goodness of creation and the abundance made possible by the Creator. To disdain what is on earth to be consumed is not purely and simply virtuous.

A Hasidic story has it that a world-denying Jew heard the call to asceticism. He thought it a part of the commandments that he must do without good food, good wine, and the company of good women and friends in general. He took no place at their festive tables; he heard no good music and did without great art. All this he did with an eye on the promise of paradise for the renouncer.

He died. He did indeed find himself in paradise. But three days later, they threw him out because he understood nothing of what was going on.

In the world of Jesus, to which more than 80 percent of the American people claim some loyalty, there are also countertexts permitting and rejoicing in consumption. Just before the death of Jesus, the Gospels have a woman pouring costly ointment—"Why was this not sold and given to the poor?"—on his feet, an act then pointed to as an enduring symbol of devotion. The Gospel writers observe Jesus at banquets in the homes of the wealthy, letting them enjoy luxuries. How do we fit this into an inclusive ethic against consuming?

Add to those two observations a third: given a market economy, one that is not likely to go away, what would happen if many *did* live literally and exclusively within a simple Jesus ethic of simplicity? What would happen to the economy and thus to the distribution of human goods that now goes on?

It was an Alec Guinness film, I believe, that depicted an English village whose parish priest began to preach Jesus' injunction to a rich young ruler: Go and sell all that you have, and give to the poor. Unfortunately, one Sunday in his sparse congregation, one of the village's parishioners, a very wealthy unmarried woman, heard the sermon, believed this messenger and the message, and did precisely that. She sold all she had and distributed it. Instantly, the economy of the town broke down. Chaos is too mild a term for what occurred thereafter.

Less ridiculously, in a recent incident that demonstrated how perplexing this issue is, an evangelical magazine that advocated low consumption and high justice invented a product line of gifts: bowls, fans, and hangings imported from the poor world. The prices were relatively high because the sellers insisted that their suppliers pay the producers of the items a living wage according to the standards of the country, whether Peru or Thailand, where they were being produced. No sooner did the line of goods hit the market than the sellers received critiques and engaged in self-examination. It was pointed out to them that not one of their products was necessary. No purchaser needed one more never-to-be-used handmade wooden bowl that was too unaesthetic to be considered a work of art or one more hanging for already over-crowded walls. The decision: to keep on selling these items despite the strictures because the paying of foreign peasants was a greater good than the consumption of their goods was evil. Religious utilitarianism or utilitarian religion decreed that a greater good for a greater number would come from this policy.

Not many years ago, preachers preached against consumerism and luxury in the pre-Christmas shopping market. Their congregations, and eventually they, too, would have to hope that the message was not taken too seriously. When stores report a 2 to 3 percent gain in Christmas sales instead of the anticipated 7 percent, shopkeepers, suppliers, and manufacturers, all those de-

pendent on and related to the products and practices, and thus
eventually the parishioners and the parson, suffer. The economy
that provides jobs and measures of security depends on overpro-
duction for overconsumption. The dilemma remains.

I may have been overgenerous to the party that derides advocates of the simple life and critics of the consumption practices
of the affluent North, particularly North America. To keep things
properly tense, let me say that there are good reasons to heed the
warnings and listen to the criticisms of those who criticize consumerism. Many of them argue that we are not in a zero-sum
game. This means that the contention that high, luxurious, and
wasteful living and consumption on the part of rich nations does
not mean that rich Americans will be consuming at the expense
of the poor. Less food on the tables of the conspicuous consumers in the rich world or worlds does not necessarily mean
more food on those in the poor world or among the poor. That
issue is too complex to be entertained in detail here.

Rather, this is to agree with all the authors in this book and
elsewhere who argue that comparing "us" with "them" or comparing "you gorgers" with "us ascetics" is not the real issue. What
is at stake is overconsumption in the first place—for three reasons: the overuse of nonrenewable resources deprives those in
the future of access to them; the environmental pollution that results from emissions into the air and waste dumped on the
ground or into the rivers is destructive; and misuse of resources
does evil to what we might call the soul, the inner being, the outlook, of those guilty of consumerism.

The religious moral systems that predominate in the affluent
world, particularly in what we call the West, caution against over-
exploitation of resources and destruction of the natural environment. In biblical traditions, especially in recent years, there has
been a recovery of the call to be responsive to the gifts of a
Creator and to be responsible for stewardship. Together, these
motifs lead people to be cautious about such exploitation.

In some symposia, one hears what I have to call romantic descriptions of the ways other religious traditions—Buddhist,
Hindu, and Native American—have handled or would handle the

issue of consuming better than have Jews, Christians, Muslims, and their spiritual cousins who are heirs of Enlightenment and producers of technology. However, such depictions often obscure from sight the luxury of the few and the wasting by the many where these other religions prevail. Each has resources for dealing with consumerism, and these promise some yield as an ever more pluralist America becomes open to them. But the record of the ways in which cultures informed by them deal with technology and its products, the free market and its options and lures, leads one quickly to the conclusion that there is no place to hide. Nowhere has *Homo economicus* produced cultures that promise to shield citizens from consumerism. Some individuals or intense and disciplined groups—for example, monks—in each religious tradition have transcended the problem, but they remain very small minorities.

Remarkably, religions have paid much attention to the third issue: what consumerism does to the consumer. Here again, the dominant figure in Western religious thinking, the Jesus of the Gospels, is clear and unambiguous, and texts are many. Jesus is heard telling the parable of Lazarus and Dives. Lazarus is the rich and indulgent man who does not notice the poor man at the gate; Lazarus consumes and is found after death in the consuming flames of Gehenna. Jesus speaks of a rich and successful farmer whose granaries are full. So he tells his soul to take his ease, secure with their goods—only to find that soul taken from him that night. He is called "Fool," and his planned consumption now is beside the point (Luke 12:16–21). "A man's life consisteth not in the abundance of his possessions" (Luke 12:15). Jesus' sayings are forthright about the choice personified as Mammon versus God: a person can choose one, but not both (Luke 16:13).

In such settings, consumerism means worshiping the gift instead of the Giver or seeing oneself as self-made and then worshiping the creator, namely the self. It means letting the world of "it" take over instead of the world of "Thou," of personality and interpersonality, in divine and human transactions. Martin Buber, the philosopher of *I and Thou,* asks properly, "Can the servant of Mammon say Thou to his money?"[6] In a King James translation

of a psalm, God gave the people what they wanted but *sent lean-
ness into their souls.* Those who do not respond to such a theological observation can easily find humanistic counterparts open to
empirical observation. Regularly, critics observe that as North
Americans get more and consume more, they are increasingly unsatisfied spiritually and thus paralyzed when it comes to making
difficult decisions and taking daring steps.

If the danger is to the soul, the self, it would follow that any
step toward a different understanding of consuming has to begin
with the resolve of the individual, in whatever culture or class.
But such individual address to questions of guilt or of salvation
from such guilt does not offer or effect all that is needed for cultural, societal, and global change. However modest the expectations of an individual, most ethicists and moral calculators insist
that a change in one's ethics to the point that more thoughtful use
of resources occurs has both intrinsic and incremental effects.
Intrinsic: it is simply the right thing to do, on its own terms.
Incremental: if enough people make a change in mind and style,
they will help produce a better culture.

The most consistent and visible philosophical and religious
critic of consumerism at the century's end has been Pope John
Paul II, who cannot be accused of having been soft, even at the
edges, on communism. His persistent comments on economic
matters have shown a consistent equipoise that leads Catholic
and non-Catholic thinkers alike to quote him, when it serves their
purpose, against the other. All the while, the pope keeps articu-
lating Catholic economic positions, roughly based on natural law
theories, that go back at least to Pope Leo XIII and his *Rerum novarum* in 1891.

In 1991, John Paul II issued *Centesimus annus,* in which he
posed Marxism against "the affluent or consumer society." Such a
society, he asserted,

> seeks to defeat Marxism on the level of pure materialism by
> showing how a free-market society can achieve a greater sat-
> isfaction of material human needs than Communism, while
> equally excluding spiritual values. In reality, while on the
> one hand it is true that this social model shows the failure of

Marxism to contribute to a humane and better society, on
the other hand, insofar as it denies an autonomous existence
and value to morality, law, culture and religion, it agrees with
Marxism in the sense that it totally reduces man to the
sphere of economics and the satisfaction of material needs.[7]

Twenty paragraphs later, the pope gets explicit:

A given culture reveals its understanding of life through the
choices it makes in production and consumption. Here *the
phenomenon of consumerism* arises. Of itself, an economic
system does not possess criteria for correctly distinguishing
new and higher forms of satisfying needs from artificial new
needs that hinder the formation of a mature personality.
Thus a great deal of educational and cultural work is urgently
needed, including the education of consumers in the responsible use of their power of choice. . . .

It is not wrong to want to live better; what is wrong is a
style of life presumed to be directed towards "having" rather
than "being." Even the decision to invest in one place rather
than another, in one productive sector rather than another,
is always *a moral and cultural choice.*[8]

Having moved to that point, we are left with the polarity observed throughout this essay. How do we do justice to the calls to
renounce, forgo, and be restrained—at the expense of existing
economies? On the other hand, how do we do justice to the need
to keep economies vital by purchasing and using many goods? As
implied before, to wander onto the battlefield called "common
ground" is to invite attack from both and be satisfying to neither.
Ecologists such as Paul Ehrlich write essays to the effect that
"balanced" or "middle-of-the-road" views will not do when it
comes to our environment and resources for the future. And free-
market economists often suggest that the search for common
ground, with its mix of ideological defense of the free market and
restraint, is also a "muddled middle" position. Is there any way to
get a handle on the horns of this dilemma, any way to negotiate
the first steps toward more productive policies and patterns than
we now have?

Instead of looking for middle and middling compromises, I

would stress a concept as old as Aristotle, as countercultural as from the world of the Mennonites, as focused as it has been by Søren Kierkegaard—and brought back with systematic emphasis by, among others, John C. Haughey, S.J., in a book called *The Holy Use of Money: Personal Finance in Light of Christian Faith*. It will be economical and will exemplify a creative use of resources if I consume one of his paragraphs and reproduce it here:

> By equipoise I mean that people (and by extension the group) have sufficient self-presence that their material and financial resources are kept in a position of means while their hearts sort out the good to be chosen, the value to be expressed, or the purpose to be pursued. Synonyms for this are singleness of heart or indifference, that is, notwithstanding one's feelings one chooses to do this or forgo that, and so forth. Equipoise is a poise of spirit that can weigh conflicting pulls and not act compulsively or addictively so that one's choices or use of things express one's deeper self, one's interiority.[9]

Pages later, Haughey reviews the theme:

> What grows by equipoise is the capacity of people to determine their lives. What shrinks is the amount of determination exercised on their lives by forces outside of them. Extending the tent poles of obedience [to the call of this ethic], therefore, in this direction enhances the possibility of a commitment of oneself in freedom and love, the stuff of which eternity is made.[10]

I prefer the term *equipoise* to *balance* because *balance* connotes the inert, whereas the *poise* of *equipoise* suggests readiness for motion.

In such passages as these, the counsel to seek balance through reason or moderation (*sophrosune*, or "enough is enough") goes back to Aristotle or, for this Jesuit, to Thomas Aquinas; the theme of "singleness of heart" preoccupied Kierkegaard; positive and creative difference was a Mennonite theme (*Gelassenheit*); the concentration on eternity comes courtesy of Haughey's Catholicism. Although in a pluralistic society not all could draw on such sources and many would find this or that detail uncongenial or not easy to transplant, they should be able to see in them an address to their lives that can by analogy be made culturally and societally productive. In all cases, however, it can be seen that personal decisions are made within complex cultural contexts, and an understanding of these contributes to comprehension of what people at a particular time and place are doing to seek meaning in life.

Many analysts today see North Americans in the "shop till you drop" context of the mall, where they turn consumerist but remain unsatisfied in respect to meaning and hence in their ability to make decisions. As ethicists and theologians Herman E. Daly and John B. Cobb Jr. typically put it:

> With the breakdown of community at all levels, human beings have become more like what the traditional model of *Homo economicus* described. Shopping has become the great national pastime. . . . On the basis of massive borrowing and massive sales of national assets, Americans have been squandering their heritage and impoverishing their children.[11]

In another book, John Haughey describes this as *pleonexia,* "a passion for more," "an insatiability for more of what I already experience or have. If I just had a little bit more, I would be happy." And Haughey hears Jesus saying in Luke 12:15, "Avoid pleonexia in all its forms."[12]

French scientist and philosopher Blaise Pascal once said that rather than seek banal middles, one should assert what is valued in two extremes and then do justice to some aspect of them both. The concept and practice of equipoise does that—though then it must be complemented in and enacted through regard for the community, in the practice of politics, and with an eye to those who, not having access to resources, do not have much to worry about by way of consumerism or consumption.

Can't Get That Extinction Crisis Out of My Mind

Stephanie Mills

THE OTHER NIGHT, I was talking with Linda Grigg, an organic farmer, about consumerism. Linda said she was kind of glad that she and her husband, Jim Moses, will never make enough money to be tempted to deeds like buying a not-altogether-pleasing million-dollar house with Lake Michigan frontage and then remodeling it to the tune of $700,000 (an actual instance of megaconsumption she'd just heard about from a contractor friend). Unlike me, Linda didn't get outraged or even righteously indignant about such plutocratic show, being wise enough to see that those lakeside mansion owners might just be doing what comes naturally to folks in their position. What I might regard as vastly ostentatious spending evidently has its own internal, if hermetic, logic. Similarly, much of my spending makes sense to me but doubtless would seem gross to a real frugality adept. What is luxury to one woman is necessity to another, and most of us define *enough* as "just a little more."

For instance, I wouldn't mind owning a little more land, if only to secure my privacy. The fact that I own any land at all is fluky. There came a windfall just when my neighbor wanted to sell thirty of his acres, so I bought them because I could. I didn't want to see the smoke from my neighbor's cabin. Other than diddling with erosion control and making the occasional gesture toward ecological restoration by planting native trees, the only thing I do with my land is inhabit it. For a few years, the cabins were kept at bay. Now, all around me, as in most amenable terrain, there's a ferocious amount of land speculation going on. In what used to be passable farm country, there are more and more houses on smaller and smaller parcels. Indeed, I'm not morally certain I couldn't be driven to sell off a few acres to keep myself fed (and I'm sure I wouldn't want to watch what the new owners would do with the land). At this stage in the ecological game, almost every form of land use is really abuse.

It has become normal to possess and dispose of land as though it were something lifeless, as though a parcel could be dealt with separately from its several contexts, from watershed to bioregion to planet. Consuming not only the sustainable harvest of the land but also its very life may be the current norm, but from the standpoint of neighbors, fellow creatures, and posterity, it's wrong. As Aldo Leopold wrote fifty years ago in *A Sand County Almanac*, "A thing is right when it tends to preserve the integrity, stability, and beauty of the biotic community. It is wrong when it tends otherwise."[1] This is the essence of Leopold's land ethic.

Leopold also observed that "perhaps the most serious obstacle impeding the evolution of a land ethic is the fact that our educational and economic system is headed away from, rather than toward, an intense consciousness of land." Of course, an intense nonpecuniary consciousness of land can be pretty depressing.

Leopold himself bleakly wrote, "One of the penalties of an ecological education is that one lives alone in a world of wounds."[2]
What some folks term development looks like mutilation to me.

I live not far from a perfect swimming hole, a sweet little lake
that was the first place I set foot in when I arrived in this county
that became my new home. After a long flight from California and
a long drive from the airport, my host took me to the lake to go
swimming by starlight. Back then, in 1984, there was a lone cottage on the lake's wooded shore whose waterskiing proprietors an-
nually paid it a brief summer visit. The rest of the year, the lake,
to which some earlier owners had granted a slice of public access,
was a neighborhood commons: an unsung Lido for families,
who'd bring their kids, who'd bring their inner tubes; for fishermen or fisherwomen with low-horsepower boats; and for the occasional translake swimmer. (Me. I confess my vested interest.)
In the shallows, there'd be minnows for the kids to chase. Once
I even heard a hidden loon cut loose with its spooky yodel.
Occasionally, a great blue heron would stalk the waterside at
dawn, preempting the fishermen's luck. The lake was, in a word,
paradise.

Then, a few years ago, at the turnoff to the lake there appeared
a billboard advertising narrow building lots pivoting like the sticks
of a Japanese fan on a wee hub of lake frontage, with a cheery
sign announcing a future subdivision. By and by, the lots began to
sell. Each sale was proclaimed on the hateful map. Every time I
bicycled past the sign on my way to the swimming hole, I'd heap
contumely on the developer. After a year of this, I decided that
this cursing was bad for my soul. I asked the developer if he
would meet with me. I'm not sure exactly what I was hoping for:
for us to talk as human beings; for me to offer amends for my reflexive hostility; for a miracle, maybe. What I got was a shocking
encounter that upbraided my innocence and reduced me to tears,
which was not all that easy to do. My weeping moved him not one
whit. When I told him I cared about what happened to all the life
around the lake, he flew into a rage. He hotly deplored the fact
that the long-ago owners of the lake had ever granted any public
access. Rightly enough, he pegged me as an environmentalist and

an Indian-lover. He more or less suggested that I and my senti-
mental ilk were bent on snatching the bread from the mouths of
his babes. Et cetera.

The subdivision proceeded, the developer's babes were fed,
and several very big houses with lawns, set back and up a re-
spectful distance from the water's edge, now intrude on the lake-
side woodland. Docks and modest vessels pertaining to those
houses punctuate the shore. There are driveways now and grassy
clearings. The effect of this new settlement is the usual: fragmentation of the landscape—a reduction in its biotic integrity,
stability, and beauty. The everyday suburban conditions—lawns,
house cats, septic systems, gaps in the forest canopy, vehicles
unpredictably crossing what used to be the nocturnal animals'
paths to the water, rank vegetation flourishing in the sunny paths
of disturbance, noise during nesting season, and additional tykes
making a pastime of catching frogs and fish, competing with the
herons—are disastrous for all but the weediest plants and animals. Regardless of our envirocredentials, we builders of new
homes (for I am one) are committing habitat destruction, and not
only where we build our homes but also in the places where our
lumber comes from. Thus, and by all the rest of our consuming,
we play our own modest roles in the earth's sixth great extinction
crisis. Owing in large measure to humankind's long, steadily
accelerating career of habitat shattering, the rate of extinction
is currently about a thousand times what is normal. That's
how fast the planet's biotic community is losing member species
these days.

Living in the heyday of North American consumerism, being
mistress of my own estate, and having my choice of cars, computers, espresso machines, blow-dryers, and designer T-shirts (assuming I can pay for, or charge, them) is no consolation. I can't
get that extinction crisis out of my mind. Extinction is not abstract in the least. It's the thousands of instances of the desola-
tion of being the last of one's kind. The birder's account of hearing what may have been a lone male cerulean warbler, say, singing
and singing and singing for a mate, unaware that there may not
be a female of the species left, within hailing distance anyway, is

well on its way to becoming a genre. As the habitat goes, so go the populations. The interior forest habitat necessary for these and other wood warblers to nest unmolested and so to persist is being shredded. Because birding is a popular pastime, we have more information about the fate of the birds of the forest than about their less conspicuous associates—salamanders, mites, shrews, beetles—but we can reasonably assume that most of them are finding it harder to get a date, too.

Life seems to want to fill in all the niches, and to do so with wondrous diversity. Through the generations, plants and animals shape themselves to place. "Place" could be as small and fleeting as the receding bands of moisture in yard-wide vernal pools, whose annual evaporations make for a quick succession of infinitesimal wildflowers and their insect associates. Bulldoze a few California acres for a housing development and those fairy gardens are gone for good. "Place" also could be the vast unbroken hardwood forest of presettlement eastern North America, whose bounty of seeds, nuts, and berries could sustain flocks of billions of passenger pigeons. Although it's an exaggeration to say we're *destroying* the earth, human activity has been simplifying and diminishing places, eliminating populations of plants and animals, disrupting intricate relationships, and laying places open to alien invasions (*alien* meaning not "from outer space" but, in this instance, "foreign to this particular ecosystem").

Thomas Ford, a local wildlife artist and birder I consulted about the wood warblers, told me the following story of alienation. The brown-headed cowbirds, thanks to the edges opened by roads and clearings, now have forest venues in which to fool other birds into rearing cowbird young. Cowbirds (once called buffalo birds) learned this trick when they followed the bison across the Great Plains, feeding on the grasshoppers kicked up by those millions of hooves. The cowbirds would deposit their eggs in any nests they could find and then roam on, leaving their offspring in foster care. Today, with fields and lawns and roads encroaching on forests everywhere, the cowbirds have access to the nests of warblers and other birds of the interior. Cowbird babies usurp the warblers' nurture—thus more cowbirds and fewer warblers et al. It's a typical weed story, one whose details could hardly have been included in the land-use plan, but that illustrates the unforeseen consequences of actions some refer to as growth, progress, and development. With millions of such disturbances, humanity's impact on the whole planet is inordinate.

An article in the July 1997 issue of *Science* quantified some aspects of this impact:

> Between one-third and one-half of the earth's land surface has been transformed by human action; the carbon dioxide concentration in the atmosphere has increased by nearly 30 percent since the beginning of the Industrial Revolution; more atmospheric nitrogen is fixed by humanity than by all natural terrestrial sources combined; more than half of all accessible surface fresh water is put to use by humanity; and about one-quarter of the bird species on earth have been driven to extinction.[3]

Changes of this magnitude weren't effected all at once, although lately they've gone into overdrive. Consumption of the very life of the land is not an American invention; it is the habit of civilization. Yet the European encounter with the Americas, begun a mere five centuries ago, was like a spark to tinder: patterns of land use that over a couple of millennia had domesticated, sometimes desertified, and definitely aged the Old World were recapitulated in decades. Civilization befell the wilderness of North America and beset a continent peopled by a few million Indians, whose diverse cultures—constituting hundreds of place-specific variations on the theme of subsistence—held them in equilibrium with the land.

Setting aside the controversial idea that the Paleo-Indians who crossed the Bering land bridge had a hand in the extinction of a mammalian megafauna—big game that would dwarf any terrestrial animal alive today—it is not idealizing indigenous peoples to say that their inhabitation of the Americas wrought only subtle changes—that is fact. Evidence of the Indians' deft tending of woodlands, prairies, and deserts is only now being recognized. Today, the globalization of North American–style consumerism threatens to swamp any remaining indigenous cultures that enjoy an ethical relationship with their habitat. Left to their own devices, indigenous peoples might eventually have invented their own ways to mass-produce doodads from the raw materials of wilderness. We'll never know. Nevertheless, we'd do well to consider the merits of the animistic worldview that fostered their vital sense of community with more-than-human nature and their capacity to experience contentment without possessions.

Subsistence or foraging peoples tended to share their few goods and to diffuse authority. Hunter-gatherers responded to food scarcity by migration or raw endurance, by having few children, or by dying in their thirties. Grain storage was not an option. Sustenance came directly from the land through a sacramental exchange; flourishing depended on utter alertness and was the occasion of gratitude. Austerity might alternate with feasting. A Plains Indian might in one sitting have eaten several pounds of buffalo meat fresh from a kill. Hungry people will gorge themselves. It may be that today's rampant consumerism betokens the soul famine of a society estranged from the living earth.

Even though North American consumerism may be the culminating stage of a lengthy historical process, that's not to say that consumerism is somehow natural. Throughout human history, the power elite have played a role in steering us away from participation in nature and toward objective materialism.

With agriculture, humanity moved a notch closer to consumerism. Farming produces the storable surplus commodities that permit fixed settlements, centralization, specialization, and social stratification, thus civilization. Ever since the earliest city-states and theocracies were formed, the power elite have been the leading exponents of the joys of consumption, obtaining their luxuries and leisure by *force majeure* or by invoking the fear of God or the myths of divine right, profit sharing, or trickle-down. In order to support the kings, khans, pharaohs, emperors, thanes, lords, and chief executive officers in the manner to which these rulers had become accustomed, and to which they claimed to be entitled, the subject populace would exploit the land.

Give or take a few exhausted bioregions (the Fertile Crescent, for example, or the once wooded Mediterranean basin), this agrarian civilization "worked" for about forty-five centuries. With the waning of the Middle Ages, trade began to break the bonds of custom, to desacralize life, and to commercialize human exchange. Land, labor, and capital became commodities. Sustenance increasingly became contingent on the ability to get money to buy rather than the ability to raise or make the basic necessities. The necessity of maintaining the land's health became less immediate or practicable to the many.

Early English emigration to North America wasn't driven so much by a love of adventure as by a massive outbreak of poverty. The wool market was booming in the sixteenth century. To pasture sheep to meet the demand, what had formerly been the common lands of villages were "enclosed," or privatized. Country people who had for centuries allocated among themselves the various, if meager, products of the commons—forage, firewood, stall litter—found themselves displaced from the land by fiat. Now they were "surplus population." Seeking but seldom finding employment and having to sell their labor cheap, they might drift into the cities to beg. Or, in exchange for passage across the Atlantic Ocean, they might indenture themselves for a stretch of years, hoping thereafter to enclose a bit of the aboriginal commons of North America. Very few nouveau poor English settlers realized such hopes, though. In the new country, property would be distributed almost as unevenly as it had been in the old.

Regardless of the colonists' estate, to a people whose idea of landscape had been formed in England's green and long-domesticated countryside, the encounter with the vast, somber eastern forest had to be terrifying. The whole seaboard was very likely appalling, having weather of shocking ferocity, being devoid of gold, and apparently being ill suited to familiar crops. These immigrants from a land long since deforested surmounted their terror and assuaged their poverty or greed in a frenzy of conquest and consumption. In America they saw wood, not forests: boundless fuel, a wealth of ships' timber, a riot of game and no royal foresters to bar the taking of any of it; a lavish invitation to hunt and hew and feast before the fire.

It was only in a limited sense a people's free-for-all. The hell-bent consumption of the American land has been largely a matter of business. Joint stock companies—forerunners of corporations—were chartered to colonize the New World and to profit their investors. The plantation agriculture that produced such commodities as tobacco, indigo, and sugar required that land be cleared and cheaply worked by indentured servants or slaves. Farther north and west, the fur trade reached, in the name of fashion, deep into the American wilderness for pelts of ermine, mink, otter, marten, fisher, wolverine, red fox, gray wolf, black bear, and beaver. It was an enormous enterprise. According to Peter Matthiessen in his *Wildlife in America,*

> A Hudson's Bay Company sale, in November 1743, disposed of 26,750 beaver pelts, as well as 14,730 martens and 1850 wolves; that these were by no means the only victims, even among their own kind, is indicated by the fact that 127,080 beaver, 30,325 martens and 1,276 wolves, as well as 12,428 otters and fishers, 110,000 raccoons, and a startling aggregation of 16,512 bears were received in the French port of Rochelle in the same year. People today who have no reasonable expectation of seeing even one of these creatures in the wild without considerable effort to do so might well look carefully at these figures.[4]

The Lakota people called the whites *wasichu,* which means "he who steals the fat." Were there also words meaning "one who fells whole hemlock groves for the tanbark, leaving the logs to rot," "one who hoses away the soil of the Sierra Nevada foothills in search of gold," "one who chops down venerable pecan trees to harvest the nuts," or "one who burns 200-pound Great Lakes sturgeon for steamboat fuel"? For all these deeds characterized certain immigrants, too.

Prior to civilization, human groups lived within their ecosystems or watersheds—they inhabited their life places. By our myriad cultures, we have been a diversely niche-adapted species. But the religion of commerce undermines loyalty to place. Yet at the beginning of the twentieth century, there still were redoubts of rural subsistence in Europe and America. Many of our great-grandparents provided their own food, clothing, shelter, transportation, and entertainment. To pay for the goods they couldn't make by themselves or do without, they raised cash crops. Excess was seldom a problem. At century's end, we're consumers, not gatherers or producers. We're at the mercy of dimly understood industrial processes and long lines of supply. Being at such removes—practical, geographic, and technological—from our sustenance, most of us are ignorant of the source of our tap water and the provenance of our food. Unless we live in Nigeria or Kuwait or on Alaska's North Slope, we lack physical intelligence about what is entailed in fueling the millions of internal combustion engines on which we utterly depend. If we're not from rural Maine, northern Ontario, the central Appalachians, or the Pacific Northwest, we're missing information about the origins of our paper supply and the costs of logging downslope. Because the sources, processing, and manufacture of our goods are so widely scattered, it's nearly impossible for us to comprehend the effects of our way of life on the biotic community. Today, not just North Americans but the wealthier portion of the world's people command their sustenance and luxuries from the biosphere as a whole. Those who have the means shop the global marketplace; those who don't, stock it.

Knowledge of some of the true costs of our consumer goods can induce a miserable consciousness of being stuck in habits that are Life (with a capital *L*) threatening. "If you eat bananas, you're in favor of low-intensity warfare in Central America," declared a young Mexican agronomist at an Iowa conference a few years ago. This is the part left out of Chiquita's story. "Consumerism is complicity," says bioregionalist Peter Berg. Although I'd like to think I'm in solidarity with campesinos everywhere, bananas don't grow in my bioregion, and I eat quite a few.

Even though I do a fairly good job of minimizing my garbage output, I'm participating nevertheless in a system that seizes and wastes on my behalf, not far enough out of sight and seldom out of mind. Because I live a mile north of a regional sanitary landfill, our waste is in my face. The dozen or so times a year I go to the dump to dispose of my bags of trash, I confront the throwaway part of consumerism. But the wads of nonbiodegradable waste are only the gross evidence of consumption: each American could fill 300 shopping bags per week with the resources she or he uses—the coal, the gas, the grain, the soil that washes away. Still, I can't conceive of not recycling, even though I know it's like bailing with a perforated bucket. Diverting a little trickle of the waste stream is a righteous gesture, an effort not to squander the earth, but a trip to the mall—headwaters of the flood of dreck that debouches at the landfill—hints that to recycle may be to approach the problem from the wrong end.

I do bear moral responsibility for the consequences of my consumption, but we consumers didn't originate the lifestyle. It has taken relentless, well-crafted persuasion—and occasional coercion—to override the common values of frugality and sharing. Over the course of the twentieth century, by means of mass production and global transportation, an ersatz version of the gluttony of nineteenth-century financiers has been democratized. The process feeds itself: as commerce reduces the beauty, abundance, and complexity of the land, nature compares ever less favorably with the bazaar. But even Muzak and designer athletic shoes can't make up for ecosystem collapse. As earth's wildness and human cultural diversity are rendered down to feedstock for the global economy, we do begin to notice. This rendering, having about gone the limit, is becoming harder to sell.

On the buyers' part, consumerism may seem mindless, but sellers take premeditated aim at spots in the ego or unconscious mind, making amoral, if often clever, symbolic equations: liquor and sex, tobacco and virility, automobiles and freedom, cosmetics and allure, soft drinks and happiness, pharmaceuticals and health, cell phones and family ties. Double-page magazine spreads of sport utility vehicles perched alone and splendid atop some red-rock mesa beckon you and the kids to be the next Lewis and Clark expedition, hassle free as long as you stay within range of a filling station. Alas, painted cakes do not satisfy hunger. We cannot buy our way out of this situation, and the market will not lead us. To arrest the final consumption of the earth, nothing short of epochal, devolutionary change of the political economy is called for.

If, say, by prodigies of conscience, frugality, cooperation, and peaceful transformation, America bends its way of life toward preserving the integrity, stability, and beauty of the biotic community—even if all of us reduce our consumption to the barest of minimums—one other great change will be necessary. Per capita consumption is a ratio. In considering such ratios, it is, of course, important to take class differences into account. People who consume their grain in the form of meat, milk, and beer, who subscribe to the daily newspaper, who have second homes and third cars and travel by jet, are devouring a greater share of the planet's substance than are vegetarian homebodies who read their news at the public library, assuming these herbivores travel on foot or by bicycle and not in ethanol-powered vehicles. Nevertheless, if the human population continues to burgeon, at some not-too-distant point even the most parsimonious lifestyles will become unsustainable. Therefore, one of the most obvious (if not one of the easiest) ways to rein in consumerism is to refrain from bringing more consumers into existence.

"By every conceivable measure," observes Edward O. Wilson in *The Diversity of Life,* "humanity is ecologically abnormal. Our species appropriates between 20 and 40 percent of the solar energy captured in organic material by land plants. There is no way that we can draw upon the resources of the planet to such a degree without drastically reducing the state of most other species."[5]

Despite the vast pollutions, extinctions, and privations we have wrought, *Homo sapiens* might persist. But will human beings?

The cure for our ravening estrangement from, and destruction of, the rest of the biotic community is *reinhabitation.* Being organisms, we might reestablish our creaturely reciprocity with our diverse habitats, our life places. Summoning our wit and will and capacity for invention, we need to develop the technologies, economies, and cultures that will allow us to dwell in our ecosystems in perpetuity. We can certainly learn enough about our life places to discern what kinds and sizes of human communities they will support without diminishing the future prospects of all our relations—furred, feathered, finned, fanged, and fungal.

As we exercise reproductive responsibility along with our rights and human populations dwindle to just proportions, as we revere and preserve what remains of free nature and wild land, as we set about restoring damaged land, we may regain our gut sense of the wonder and revelation in more-than-human life. It's time to ditch the home entertainment center and break the consumer trance, time to roll up our sleeves and learn the plants.

We have not made ourselves; we were made by the necessities of foraging and flight, informed by stories told and told again around the wandering band's campfire, by movement with the seasons. Human beings coevolved with forests and savannas, with rivers teeming with fish and skies blackened by fowl; we came into being in the company of thundering herds of ungulates and the omnipresent threat of pitiless predators.

Extinction is forever, but where there are life-forms, there is hope. Enough biological diversity may remain that we can reanimate our landscapes. We may even rescue the wildness within us from the extinction threatened by credit cards, muscle wagons, and trips to the mall. By working to restore our life places from the soil on up, we can renew our membership in the biotic community.

The good life, incarnate, is a birthright. It's about being and doing, not having. "Every cup of water is a prayer," says a friend who pumps his own from a hand-dug well. Every trip to the swimming hole is a pilgrimage, a baptism in the waters of life. And the nature of life is *enough*.

Notes

Introduction

by Roger Rosenblatt

1. Thorstein Veblen, *The Theory of the Leisure Class* (1899; reprint, New York: Penguin Books, 1967).

2. Speech given by Hillary Rodham Clinton at the annual meeting of the World Economic Forum, Davos, Switzerland, February 2, 1998 (Internet: <http://docs.whitehouse.gov/WH/EOP/First_Lady/html/generalspeeches/1998/19980202.html>, October 31, 1998).

3. John Galsworthy, *The Man of Property* (London: Heinemann, 1911), p. 41.

4. Yiannis Gabriel and Tim Lang, *The Unmanageable Consumer: Contemporary Consumption and Its Fragmentation* (Newbury Park, Calif.: Sage Publications, 1995); Jackson Lears, *Fables of Abundance: A Cultural History of Advertising in America* (New York: Basic Books, 1995).

5. Robert Heilbroner, *Visions of the Future: The Distant Past, Yesterday,*

Today, Tomorrow (Oxford, England: Oxford University Press, 1995), p. 54.

One World of Consumers
by William Greider

1. Herman E. Daly and John B. Cobb Jr., *For the Common Good: Redirecting the Economy toward Community, the Environment, and a Sustainable Future* (Boston: Beacon Press, 1994).

What's Wrong with Consumer Society?
Competitive Spending and the "New Consumerism"
by Juliet Schor

1. Thorstein Veblen, *The Theory of the Leisure Class* (1899; reprint, New York: Penguin Books, 1967).

2. Pierre Bourdieu, *Distinction: A Social Critique of the Judgement of Taste* (Cambridge, Mass.: Harvard University Press, 1984), pp. 1–2.

3. Ibid.

4. W. Lloyd Warner, Marchia Meeker, and Kenneth Eells, *Social Class in America: A Manual of Procedure for the Measurement of Social Status* (New York: Harper, 1960).

5. Douglas Holt, "Does Social Class Structure Consumption?" *Journal of Consumer Research* 25 (June 1998).

6. James S. Duesenberry, *Income, Saving, and the Theory of Consumer Behavior* (Cambridge, Mass.: Harvard University Press, 1949); Robert H. Frank, *Choosing the Right Pond: Human Behavior and the Quest for Status* (New York: Oxford University Press, 1985); Robert H. Frank, "The Demand for Unobservable and Other Nonpositional Goods," *American Economic Review* 75, no. 1 (1985): 101–116.

7. Veblen, *Theory of the Leisure Class*.

8. Juliet B. Schor, "Do Americans Keep Up with the Joneses? The Impact of Consumption Aspirations on Savings" (unpublished paper, Tilburg University, Tilburg, North Brabant, Netherlands, 1997). See also Juliet B. Schor, *The Overspent American: Upscaling, Downshifting, and the New Consumer* (New York: Basic Books, 1998).

9. Susan Fournier and Michael Guiry, "A Look into the World of Consumption Dreams, Fantasies, and Aspirations." Research Report. University of Florida (December 1991), p. 15.

10. Thomas C. O'Guinn and L. J. Shrum, "The Role of Television in the Construction of Consumer Reality," *Journal of Consumer Research* 23, no. 4 (1997): 278–294. Quote from p. 279.

11. L. J. Shrum et al., "Processes and Effects in the Construction of Normative Consumer Beliefs." *Consumer Research* 18 (1991): 755–763.

12. Juliet B. Schor, "Do Americans Keep Up with the Joneses?," Table 4.2 and pp. 77–83.

13. Merck Family Fund, "Yearning for Balance" Poll, February 1995, data provided to author.

14. Fournier and Guiry, "A Look into the World," pp. 16–17.

15. Juliet B. Schor, 1998, "Do Americans Keep Up with the Joneses?," pp. 16–17, Table 1.3 and Table 1.4.

16. Ibid, p. 15, Table 1.2.

17. James Brooke, "Who Braves Piranha Water? Your Avon Lady!" *New York Times,* July 7, 1995, p. A4.

Other Reading

Bearden, William O., and Michael J. Etzel. "Reference Group Influence and Product and Brand Purchase Decisions." *Journal of Consumer Research* 9 (1982): 183–194.

Belk, Russell. "Possessions and the Extended Self." *Journal of Consumer Research* 15 (September 1988): 139–168.

Brown, Clair. *The Standard of Living.* Cambridge, Mass.: Blackwell, 1994.

Chao, Angela, and Juliet B. Schor. "Empirical Tests of Status Consumption: Evidence from Women's Cosmetics." *Journal of Economic Psychology* 19, no. 1 (1998): 101–131.

Chapin, F. Stuart. "A Measurement of Social Status." Chapter 19 in *Contemporary American Institutions: A Sociological Analysis,* pp. 373–397. New York: Harper & Brothers, 1935.

Childers, Terry L., and Akshay R. Rao. "The Influence of Familial and Peer-Based Reference Groups on Consumer Decisions." *Journal of Consumer Research* 19, no. 2 (1992): 198–211.

Clark, Andrew E., and Andrew J. Oswald. "Unhappiness and Unemployment." *Economic Journal* 104 (May 1994): 648–659.

Congleton, Roger. "Efficient Status Seeking: Externalities and the Evolution of Status Games." *Journal of Economic Behavior and Organization* 11 (1989): 175–190.

Deaton, Angus. *Understanding Consumption.* Oxford, England: Clarendon Press, 1992.

Diener, Ed, et al. "The Relationship between Income and Subjective Well-Being: Relative or Absolute?" *Social Indicators Research* 28 (1993): 195–223.

Easterlin, Richard A. "Does Money Buy Happiness?" *Public Interest* 30 (winter 1973): 3–10.

———. "Will Raising the Incomes of All Increase the Happiness of All?" *Journal of Economic Behavior and Organization* 27 (1995): 35–47.

Festinger, Leon. "A Theory of Social Comparison Processes." *Human Relations* 7, no. 2 (1954): 117–140.

Godbey, Geoff, and John Robinson. *Time for Life.* University Park: Pennsylvania State University Press, 1997.

Halle, David. *Inside Culture: Art and Class in the American Home.* Chicago: University of Chicago Press, 1993.

Hirsch, Fred. *Social Limits to Growth.* Cambridge, Mass.: Harvard University Press, 1976.

Holman, Rebecca. "Product Use As Communication: A Fresh Appraisal of a Venerable Topic." In *Review of Marketing,* edited by B. M. Enis and K. J. Roering, pp. 106–119. Chicago: American Marketing Association, 1981.

Inglehart, Ronald, and Jacques-Rene Rabier. "Aspirations Adapt to Situations—but Why Are the Belgians So Much Happier Than the French?" In *Research on the Quality of Life,* edited by Frank M. Andrews, pp. 1–56. Ann Arbor: University of Michigan, Institute for Social Research, Survey Research Center, 1986.

James, Jeffrey. *Consumption and Development.* London: Macmillan, 1993.

———. "Positional Goods, Conspicuous Consumption, and the International Demonstration Effect Reconsidered." *World Development* 15, no. 4 (1987): 449–462.

Lane, Robert. *The Market Experience.* Cambridge, England: Cambridge University Press, 1991.

————. "The Road Not Taken: Friendship, Consumerism, and Happiness." *Critical Review* 8, no. 4 (1994): 521–554.

Leach, William. *Land of Desire: Merchants, Power, and the Rise of a New American Culture.* New York: Pantheon Books, 1993.

Lears, T. J. Jackson. "From Salvation to Realization: Advertising and the Therapeutic Roots of the Consumer Culture." In *The Culture of Consumption,* edited by Richard Wightman and T. J. Jackson Lears, 1–38. New York: Pantheon Books, 1983.

Lebergott, Stanley. *Pursuing Happiness: American Consumers in the Twentieth Century.* Princeton, N.J.: Princeton University Press, 1993.

McAdams, Richard H. Pages 1–104 in *"Relative Preferences": The Yale Law Journal.* New Haven, Conn.: Yale Law Journal Company, 1992.

MacCannell, Dean. *The Tourist: A New Theory of the Leisure Class.* New York: Schocken Books, 1989.

McCracken, Grant. *Culture and Consumption: New Approaches to the Symbolic Character of Consumer Goods and Activities.* Bloomington: Indiana University Press, 1990.

Neumark, David, and Andrew Postlewaite. "Relative Income Concerns and the Rise in Married Women's Employment." NBER Working Paper No. 5044. Cambridge, Mass.: National Bureau of Economic Research, 1995.

Park, C. Whan, and V. Parker Lessing. "Students and Housewives: Differences in Susceptibility to Reference Group Influences." *Journal of Consumer Research* 4 (1977): 102–110.

Rauscher, Michael. "Demand for Social Status and the Dynamics of Consumer Behavior." *Journal of Socio-Economics* 22 (1993): 105–113.

Schor, Juliet B. "Can the North Stop Consumption Growth? Escaping the Cycle of Work and Spend." In *The North, the South, and the Environment,* edited by V. Bhaskar and Andrew Glyn, pp. 68–84. London: Earthscan, 1995.

————. *The Overspent American: Upscaling, Downshifting, and the New Consumer.* New York: Basic Books, 1998.

————. *The Overworked American: The Unexpected Decline of Leisure.* New York: Basic Books, 1992.

Veenhoven, Ruut. "Is Happiness Relative?" *Social Indicators Research* 24 (1991): 1–24.

False Connections

by Alex Kotlowitz

1. Sarah Young, telephone interview with the author, January 1998.
2. Ibid.
3. Jonathan Kozol, *Amazing Grace: The Lives of Children and the Conscience of a Nation* (New York: HarperPerennial, 1996).
4. Joshua Levine, "Baad Sells," *Forbes* 159, no. 8 (April 21, 1997): 142.
5. Sarah Young, telephone interview with the author, January 1998.

Oh, Isaac, Oh, Bernard, Oh, Mohan

by Bharati Mukherjee

1. Wallace Stegner, *Where the Bluebird Sings to the Lemonade Springs: Living and Writing in the West* (New York: Random House, 1992), p. 10.
2. Satya Sreenivas, "Microsoft Acquires Hotmail," *India Currents* (February 1998): 11, no. 11: 12.
3. Statistics are from Ronald T. Takaki, *Strangers from a Different Shore: A History of Asian Americans* (New York: Penguin Books, 1990), 294.

A News Consumer's Bill of Rights

by Suzanne Braun Levine

1. Christine Todd Whitman, in a speech to the American Society of Magazine Editors, August 25, 1997.

Movies and the Selling of Desire

by Molly Haskell

1. See, for example, Vance Packard, *The Hidden Persuaders* (New York: McKay, 1957).
2. Gertrude Himmelfarb, *The De-moralization of Society: From Victorian Virtues to Modern Values* (New York: Knopf, 1995).
3. Olive Higgins Prouty, *Stella Dallas: A Novel* (Boston: Houghton Mifflin, 1923).
4. Thomas C. Frank, *The Conquest of Cool: Business Culture, Counterculture, and the Rise of Hip Consumerism* (Chicago: University of Chicago Press, 1997).

NOTES   213

The Ecology of Giving and Consuming
by David W. Orr

1. N. Myers, "Consumption: Challenge to Sustainable Development," *Science* 276 (April 4, 1997): 53–57; Jeffery Vincent and Theodore Panayotou, " . . . Or Distraction?" *Science* 276 (April 4, 1997): 53–57; M. Sagoff, "Do We Consume Too Much?" *Atlantic Monthly* (June 1997): 80–96.

2. Thorstein Veblen, *The Theory of the Leisure Class* (1899; reprint, Boston: Houghton Mifflin, 1973); Stuart Ewen, *Captains of Consciousness: Advertising and the Social Roots of the Consumer Culture* (New York: McGraw-Hill, 1976); William R. Leach, *Land of Desire: Merchants, Power, and the Rise of a New American Culture* (New York: Pantheon Books, 1993).

3. Paul Hawken, "Natural Capitalism," *Mother Jones* (April 1997): 44.

4. Clifford Cobb et al., "If the GDP Is Up, Why Is America Down?" *Atlantic Monthly* (October 1995): 2–15.

5. Paul L. Wachtel, *The Poverty of Affluence: A Psychological Portrait of the American Way of Life* (New York: Free Press, 1983), p. 65.

6. William Rees and Mathis Wackernagel, *Our Ecological Footprint: Reducing Human Impact on the Earth* (Philadelphia: New Society, 1995).

7. Dan Fagin and Marianne Lavelle, *Toxic Deception: How the Chemical Industry Manipulates Science, Bends the Law, and Endangers Your Health* (Secaucus, N.J.: Birch Lane Press, 1996).

8. Wendell Berry, *The Gift of Good Land* (San Francisco: North Point Press, 1981) p. 281.

9. Henry David Thoreau, *The Portable Thoreau*, Carl Pope, ed. (New York: Viking Press, 1964), p. 286; Aldo Leopold, *A Sand County Almanac* (New York: Ballantine, 1996), p. 190.

10. Janine Benyus, *Biomimicry* (New York: Morrow, 1997); John Tillman Lyle, *Regenerative Design for Sustainable Development* (New York: Wiley, 1994); Sim Van der Ryn and Stuart Cowan, *Ecological Design* (Washington, D.C.: Island Press, 1995); David Wann, *Biologic* (Boulder, Colo.: Johnson Books, 1990).

11. Vaclav Havel, *Living in Truth* (London: Faber & Faber, 1987).

12. See J. Jacobs, *The Death and Life of Great American Cities* (New York: Vintage, 1961).

13. Jacquetta Hawkes, *A Land* (New York: Random House, 1950), p. 202.

14. Ivan D. Illich, *Energy and Equity* (New York: HarperPerennial, 1974).

15. Michael J. Kinsley, *Economic Renewal Guide: A Collaborative Process for Sustainable Community Development* (Snowmass, Colo.: Rocky Mountain Institute, 1997).

16. William Ophuls and A. Stephen Boyan, *Ecology and the Politics of Scarcity Revisited: The Unraveling of the American Dream* (New York: Freeman, 1992), p. 288.

17. M. Gadgil et al., "Indigenous Knowledge for Biodiversity Conservation," *Ambio* (May 1993): 151–156.

It All Begins with Housework
by Jane Smiley

1. Morris Birkbeck, "Notes on a Journey in America from the Coast of Virginia to the Territory of Illinois," 43 (Piccadilly: Ridgeway, 1818), MSA SC 1399-1-433, Special Collections, Maryland State Archives, Annapolis, Md.

2. Catharine Esther Beecher, *A Treatise on Domestic Economy, for the Use of Young Ladies at Home, and at School* (Boston: Marsh, Capen, Lyon, and Webb, 1841), p. 5.

3. Susan Strasser, *Never Done: A History of American Housework* (New York: Pantheon Books, 1982), p. 36.

4. Ibid., p. 57.

5. Harriet Beecher Stowe, *Uncle Tom's Cabin* (New York: Harper & Row, 1965), p. 160.

6. Ibid., p. 158.

7. Ibid., p. 161.

8. Ibid.

9. Charles Dickens, *American Notes* (New York: Random House, 1996), p. 184.

Equipoise
by Martin E. Marty

1. Alasdair MacIntyre, *A Short History of Ethics* (New York: Macmillan, 1996), p. 183.

2. David R. Loy, "The Religion of the Market, "*Journal of the American Academy of Religion* 65(2): 278.

3. Andrew Carnegie, *The Gospel of Wealth* (Cambridge, Mass.: Applewood Books, 1889).

4. Adam Smith, *An Inquiry into the Nature and Causes of the Wealth of Nations* (New York: Regnery, 1998), p. 14.

5. Alexis de Tocqueville, *Democracy in America* (London: Oxford University Press, 1947), 334, 312, 311, 313); see John C. Haughey, S.J., *The Holy Use of Money: Personal Finance in Light of Christian Faith* (Garden City, N.Y.: Doubleday, 1986), p. 98.

6. Quoted in Timothy Gorringe, *Capital and the Kingdom: Theological Ethics and Economic Order* (Maryknoll, N.Y.: Orbis Books, 1994), p. 29.

7. Pope John Paul II, *Centesimus annus* May 1, 1991, par. 19.

8. Ibid., par. 39.

9. John C. Haughey, S.J., *The Holy Use of Money*, p. 142.

10. Ibid., p. 145.

11. Herman E. Daly and John B. Cobb Jr., *For the Common Good: Redirecting the Economy Toward Community, the Environnment, and a Sustainable Future* (Boston: Beacon Press, 1994), p. 187.

12. John C. Haughey, S.J., *Virtue and Affluence: The Challenge of Wealth* Kansas City, Mo.: Sheed and Ward, 1997), p. 19.

Can't Get That Extinction Crisis Out of My Mind
by Stephanie Mills

1. Aldo Leopold, *A Sand County Almanac, with Other Essays on Conservation from Round River* (New York: Ballantine Books, 1970), p. 262.

2. Ibid., p. 261.

3. Peter M. Vitousek, Harold A. Mooney, Jane Lubchenco, and Jerry Melillo, "Human Domination of Earth's Ecosystems." *Science* 277 (July 25, 1997): 494.

4. Peter Matthiessen, *Wildlife in America* (New York: Viking Press, 1959, revised and updated in 1987 by Viking Peguin), p. 81.

5. Edward O. Wilson, *The Diversity of Life* (Cambridge, Mass.: The Belknap Press of the Harvard University Press, 1992), p. 272.

About the Authors

WILLIAM GREIDER is the author of *One World, Ready or Not: The Manic Logic of Global Capitalism* (Simon & Schuster, 1997); *Who Will Tell the People: The Betrayal of American Democracy* (Simon & Schuster, 1992); and other books. He writes regularly on national affairs for *Rolling Stone* magazine.

MOLLY HASKELL has taught writing and film at Barnard College and Columbia University. Her books include *From Reverence to Rape: The Treatment of Women in the Movies* (Holt, Rinehart & Winston, 1973, revised and updated in a 1989 University of Chicago Press edition); *Love and other Infectious Diseases: A Memoir* (Citadel Press, 1992); and *Holding My Own in No Man's Land: Women and Men and Films and Feminists* (Oxford University Press, 1997).

ALEX KOTLOWITZ is the author of *The Other Side of the River: A Story of Two Towns, a Death, and America's Dilemma* (Nan A. Talese/Doubleday, 1998) and *There Are No Children Here: The Story of Two Boys Growing Up in the Other America* (Nan A. Talese/Doubleday, 1991). He frequently writes and speaks on issues concerning race and poverty. A former staff writer at the *Wall Street Journal,* he has also contributed to the *New York Times,* the *MacNeil/Lehrer NewsHour,* and National Public Radio.

SUZANNE BRAUN LEVINE is former editor of the *Columbia Journalism Review.* Previously, she was editor of *Ms.* magazine (1972–1989). She was a 1997–1998 fellow at the Freedom Forum Media Studies Center and is currently writing a book about fathering.

EDWARD LUTTWAK is senior fellow at the Center for Strategic and International Studies. He has consulted for the Office of the Secretary of Defense, the National Security Council, and the U.S. Department of State. His books include *The Endangered American Dream* (Simon & Schuster, 1993) and *Coup D'État* (Harvard University Press, 1985).

BILL McKIBBEN'S books include *Maybe One: A Personal and Environmental Argument for Single-Child Families* (Simon & Schuster, 1998); *Hope, Human and Wild* (Little, Brown, 1995); and *The End of Nature* (Random House, 1989). He has written for the *New Yorker, Outside,* and *Rolling Stone,* among numerous other publications.

MARTIN E. MARTY is director of the Public Religion Project at the University of Chicago. His numerous books include *The One and the Many: America's Struggle for the Common Good* (Harvard University Press, 1997) and *Righteous Empire: The Protestant Experience in America* (Dial, 1970), winner of the National Book Award.

STEPHANIE MILLS is the author of *Whatever Happened to Ecology?* (Sierra Club Books, 1989) and *In Service of the Wild: Restoring and Reinhabiting Damaged Land* (Beacon Press, 1995). Her articles have appeared in *Sierra, Utne Reader, Glamour,* and the *1998 Britannica Book of the Year* (Encyclopædia Britannica), among many other publications. She is currently writing *Epicurean Simplicity,* a book to be published by Island Press.

BHARATI MUKHERJEE is professor of English at the University of California, Berkeley. Her works of fiction include *Leave It to*

Me (Knopf, 1997); *Jasmine* (Grove, 1989); and *The Middleman and Other Stories* (Grove, 1988).

DAVID ORR is professor and chair of the Environmental Studies Program at Oberlin College. He is the author of *Earth in Mind: On Education, Environment, and the Human Prospect* (Island Press, 1994) and *Ecological Literacy: Education and the Transition to a Postmodern World* (State University of New York Press, 1992), as well as numerous published articles.

ROGER ROSENBLATT is an essayist for *Time* and the *NewsHour with Jim Lehrer* on PBS. His books include *Children of War* (Anchor Press/Doubleday, 1983) and *Coming Apart: A Memoir of the Harvard Wars of 1969* (Little, Brown, 1997).

ANDRÉ SCHIFFRIN is director and editor-in-chief of The New Press. He has written for the *New York Times Book Review,* the *Nation,* and the *New Republic,* as well as for scholarly journals and magazines in Great Britain, Spain, and Norway.

JULIET SCHOR is director of studies and senior lecturer on women's studies at Harvard University. Her books include *The Overspent American: Upscaling, Downshifting, and the New Consumer* (Basic Books, 1998) and *The Overworked American: The Unexpected Decline of Leisure* (Basic Books, 1992).

JANE SMILEY is the Pulitzer Prize–winning author of *A Thousand Acres* (Knopf, 1991). Her works of fiction also include *The All-True Travels and Adventures of Lidie Newton* (Knopf, 1998) and *Moo* (Random House, 1995). She is a fellow of the American Academy of Arts and Sciences.

Index

Europe:
  class systems, 16, 164
  financial crisis, global, 32
  global consumerism, 48
  savings, 52
  topography, 164
Excess as defining characteristic of the
    modern economy, 146
Expectations, consumer, 15
Expenditures and investments,
    government, 62
Expenditures for goods, personal
    consumption, 61
Extended-family ties, 55–56
Extinction crisis, sixth great, 11–12, 17,
    196–97

Fables of Abundance (Lears), 17, 18–19
Fairness and the news media, 107–8
False connections:
  brand-name goods, 10, 69
  commercialism prolonging myths, 67
  Hush Puppies, 68
  larger community, poor black children
    claiming membership to, 71
  racial divide hidden by trademarks, 72
  style, control over, 70
  white teens' romanticization of urban
    poverty, 70–71
Family, the:
  Asian immigrants, 80
  breakdown of, 61–62
  businesses run by, 7
  consuming as an addictive substitute for, 6
  emotional support, lack of, 55–56
Farming, 199–201
Farrow, Mia, 128, 129
Fashion, 10
  see False connections
Faulkner, William, 114
Federal Bureau of Investigation (FBI), 102
Federalist Papers, 165
Female, ideal viewer/consumer as, 9
  see also Housework; Women
Feminism, 10, 161–63
Feuerbach, Ludwig, 175
Fila, 68
Filene, Lincoln, 141
Fisher, Amy, 98
Flower, Richard, 155, 156
Food and lack of emotional support, 59
Food Lion, 106
Foraging peoples, 199
Ford, Thomas, 197
Forests, old-growth, 28
Forever Amber (Winsor), 115
Formaldehyde, 143
Formula feeding, 48
For the Common Good, (Daly & Cobb, Jr.),
    34

Foucault, Michel, 117
Fournier, Susan, 44
France, 52, 123
Frank, Robert, 42
Freedom, 166
Free Press, The, 119
Free will, 6
Freud, Sigmund, 131
Functional characteristics of goods and
    services, 38
Funerals, 56

Galbraith, John K., 18
Gap, the, 4
Garage sales, 143
Gardner, Erle S., 113
Gellner, Ernest, 177–78
General Motors Corporation, 142
George, Henry, 112
Germany, 52, 113
Gifford, Kathie Lee, 19–20
Gigantism, corporate, 104–6
Global consumerism:
  America accepting burden of its
    historical position, 27–28, 31–32
  American artifacts, ubiquitous nature of,
    25
  aspirational gap, 26–27, 48–49
  automobiles in China, 29–30
  crisis, financial, 32
  environmental issues, 12
  fallacy of, central, 30
  improvement, universal yearning for
    material, 31
  indigenous peoples, 198–99
  industrialization and mass consumption
    fueling, 28
  Kyoto Conference on Climate Change in
    1997, 27, 33
  mimicking American prosperity, 14–15
  new consumerism, 47–49
  political realities, 32–33
  self, confrontation with, 23
  self-knowledge advanced through, 28
  Thailand, 25–26
  transformation, economic, 33–36
  transportation, 203
  Zonis, Marvin, 173
God, belief in, 182
  see also Religion
Goldfinger, 133
Goleman, Daniel, 108
Good life and material goods, 45
Gospel of Wealth, The, 179
Gospel traditions, 14
Government changing national priorities,
    35–36, 204–5
Government expenditure and investment,
    62
Gracious Home, 4

Race, *see* Asian immigrants; Canadians,
    South Asian; Minorities
Ralph Lauren, 69
Random House, 7, 115, 118
Rap artists pushing brand names, 69–70
Recycling, 203
Reebok, 68
Reference groups, 41–44
Regulations, government, 63
Reinhabitation, 204–5
Religion:
    anti-consumerist preaching, 185–86
    asceticism, 184, 185
    colonial and biblical heritage, 14, 178
    communal economy, 180–81
    conflicts between fellow religionists,
        182–83
    consumer, what consumerism does to
        the, 187–88
    criticisms of consumerism, 37
    cults, 57–58
    eroding constraints on consumption, 47
    God, belief in, 182
    gospel traditions, 14
    Hebrew prophets, 179, 181–82
    *Holy Use of Money: Personal Finance in
        Light of Christian Faith,* 190–91
    human concerns put at top of earthly
        hierarchy, 164
    Jesus, 179–80, 185, 187
    lonely, attracting the, 58
    news media, 109
    overexploitation, cautions against, 186
    Pope John Paul II, 188–89
    Protestant ethic, 18, 177
    romantic descriptions of religious
        traditions, 186–87
    spiritual reconsideration, 2
Rerum novarum (Leo XIII), 188
Research, market, 39–40
Residential classification schemes, 40
Restraint, 11
Rhode Island, 84
Rich as a lifestyle target, the, 45–49
Robots, social, 34
Rockefeller, John D., 179
Rural landscapes, 144
Rural subsistence, 202

Sanctuary (Faulkner), 114
Sand County Almanac, A (Leopold), 194
Savings, 49, 50, 52, 63
Science, 198
Scientific pretensions surrounding market
    economics, 34
Self:
    Asian immigrants and self-fulfillment, 82
    authentic, 18
    -censorship, 105
    cultivation of, 18–19

global consumerism, 23, 28
    -interest, 180
    search of the, 20
    Western concept of, 6–7, 164
Self-defense, us-versus-them mode of, 13
Servant problem, 155–56
    see also Housework
Sherwin Williams Company, 4
Shoes, 68
Shoplifting, 47
Shrum, L. J., 44
Siblings, 56
Sikhs, 79, 80
Simplicity, shift toward voluntary, 92
Simpson, O. J., 20, 98
60 Minutes, 106
Slavery, 160–61
Slone, Alfred, 141
Smiley, Jane, 9–10
Smith, Adam, 146, 178
Sneakers, 68
Social approach to analyzing patterns of
    consumption, 38–43, 45
Social contracts, 14
Social Darwinism, 14, 179
Social relationships, disappearance of
    traditional, 41
Social robots, 34
"Song of the Shirt, The" (Hood), 19–20
Soul famine, 6, 11, 188
Spain, 62
Spending:
    census block and household, 40
    class systems, 39
    social and symbolic functions of, 38
    see also Competitive consumption;
        individual subject headings
Spillane, Mickey, 115
Spiritual reconsideration, 2
Sponsorship and athletics, corporate, 4
Sports, 4, 56
Sport utility vehicles, 30
Springer, Jerry, 106
Stanwyck, Barbara, 128, 129
Starbucks, 4
Status, 16, 41
    see also Competitive consumption
Stegner, Wallace, 78–79
Steinem, Gloria, 109
Stella Dallas, 128–30
Straight-to-the-gut format of news
    reporting, 108–10
Studs Lonigan (Farrell), 114
Subliminal advertising, 124
Subprime lending, 53
Suburbia, white, 10
    see also False connections
Sumptuary laws, 40, 41
Sustainable society, 140–43
    see also Ecology of giving and consuming